It's Not Easy
Bein' Me

It's Not Easy Bein' Me

A Lifetime of No Respect
but Plenty of Sex and Drugs

Rodney Dangerfield

HarperEntertainment

An Imprint of HarperCollins*Publishers*

FIRST EDITION

Designed by Timothy Shaner

Printed on acid-free paper

Library of Congress Cataloging-in-Publication Data has been applied for.

ISBN 0-06-621107-7

04 05 06 07 08 ❖/RRD 10 9 8 7 6 5 4 3 2

*I dedicate this book to my wife, Joan,
and to all the girls who let me sleep over*

Contents

Foreword
by Jim Carrey

The book you are holding in your hands—or clenched in your teeth, maybe?—is the amazing life story of one of my all-time heroes, Rodney Dangerfield.

I've read it twice—the first time, quickly, to see what he said about me, the second time to learn about his amazing life.

Rodney is, without a doubt, as funny as a carbon-based life-form can be. Watching his act is like watching a boxing match on fast-forward. His biggest problem is that he fires off his brilliant one-liners so fast that by the time you've recovered from one joke, you've already missed the next three. Rodney is a walking encyclopedia of stand-up comedy, spanning the generations, from nightclubs to websites, from Ed Sullivan to Conan O'Brien. And through it all, for more than fifty years, he has remained high, I mean really hip.

Foreword

In addition to performing his own comedy, he has given a big boost to hundreds of comics. As the owner of Dangerfield's, his nightclub in New York, and through his HBO specials, he has always been a young comedian's best friend. His eye for talent is unmatched, and he never took the safe way out. He fostered plenty of mainstream comedians, but his heart really went out to the edgy performers, those men (Sam Kinison) and women (Roseanne) who had a hard time getting booked when they were starting out because they weren't "user-friendly." He even helped discover a young impressionist from Canada who dreamed, at one time, of being the next Rich Little. (For those of you who are moving your lips as you read this, that young impressionist was me.) More than twenty years ago, I was performing in a small club in Toronto when I got my first gig opening for Rodney at Caesars Palace in Las Vegas. That was a very big deal for me, a huge break, and someday, I'm going to thank Rodney for giving me that break. Someday.

After that run in Vegas, Rodney took me on tour with him for a couple of years, and we had a lot of laughs, and a lot of bad airplane meals. One day, though, I decided to change my act—I wanted to stop doing my impressions and start being myself onstage. Well, things got pretty weird for a while after that. And by "weird" I mean that I was bombing night after night. But I stuck with it, mainly because I could always hear Rodney laughing in the wings. After a show, he'd say to me, "Man, those people were lookin' at you like you were from another planet!" But I was making him laugh, so I knew I was onto something. A lot of comedians, even a star as big as Rodney Dangerfield, would have dumped an opening act that wasn't making his audience laugh, but Rodney stood by me, told me to keep on doing what I was doing.

He, of course, knew something about sticking with it. He struggled for decades before he reached the top of his profession. I don't know if anybody remembers the era of the comedy club—they were quite popular places at one time, but you can only see them now in the Smithsonian, I think—but I did stand-up in clubs for fifteen years and sometimes the only thing that kept me going was the thought that Rodney had dropped out of the business when he was thirty but had come back and made it when he was in his forties. Made it big. In a business that almost always values youth over talent, he was—and still is—absolute proof that it's never too late to make your mark. You may have to quit for a while and sell some aluminum siding, but you don't have to give up your dreams.

Most people don't know this about Rodney, but he is also a very sweet and generous man. We're talking about a guy who has dozens of people walk up to him every day of his life and say, "Hey, Rodney, I'll give you some respect," as if he's never heard it before, and not once has he cold-cocked anyone. That alone is an incredible achievement. I know because, apparently, I'm smokin'!

Rodney has written thousands of great jokes, but for

me, his funniest line is his classic setup, "I don't get no respect." That's almost an inside joke because from me, and from all the hundreds of comedians he has helped and inspired, and from anybody who digs great comedy, he gets nothing but love and respect.

It's Not Easy
Bein' Me

Introduction

Here I am, eighty-two years old, writing a book. According to statistics about men in their eighties, only one out of a hundred makes it to ninety. With odds like that, I'm writing very fast. I want to get it all done. I mean, I'm not a kid anymore, I'm getting old. The other night, I was driving, I had an accident. I was arrested for hit-and-walk.

I know I'm getting old, are you kiddin'? I got no sex life. This morning, when I woke up, vultures were circling my crotch.

Hey, you know when you're really old? When your testicles tell you it's time to mow the lawn.

It's hard for me to accept the fact that soon my life will be over. No more Super Bowls. No more Chinese food. No more sex. And the big one, no more smoking pot.

Many years ago, my wife and I were living with a

friend of mine in Englewood, New Jersey. He had a big house, and we all shared it for a while.

One night I came home late and I was hungry. I saw on the kitchen table a big, beautiful German chocolate cake. Right away, the plan hit me. I smoked a joint and then I started drinking skim milk and eating chocolate cake.

Before I knew it, I had eaten half the cake.

I lit a cigarette, sat back, and relaxed.

I looked over at the remaining cake. I noticed the chocolate was moving. I didn't believe it. I looked closer. I saw there were thousands of red ants stacked at the bottom of the cake, crawling all over. But there were no ants on the side of the plate where I had eaten the cake. I knew the ants hadn't stopped at that imaginary line.

I realized I had eaten an army of red ants.

I called the hospital. They told me not to worry, it would all come out as waste. Funny. That's what a lot of people told my mother when she was pregnant with me.

Jump forward ten years. I'm forty, broke. My mother is dying of cancer. I owe $20,000 to an aluminum-siding company. My wife is sick. I've got two kids. I need money now. What am I gonna do?

Hey, wait a minute.

To tell this story right, I gotta go *way* back.

Chapter **One**
I Was a Male Hooker . . .

Most kids never live up
to their baby pictures.

Roy and Arthur was a vaudeville comedy team. Roy was my father; Arthur was my uncle Bunk. On November 22, 1921, after their last show that night in Philadelphia, Phil Roy got a call backstage, where he was told, "It's a boy!"

My father drove that night from Philadelphia to Babylon, Long Island, to greet his new son, Jacob Cohen. Me. (My father's real name was Phillip Cohen; his stage name was Phil Roy.)

I was born in an eighteen-room house owned by my mother's sister Rose and her husband. After a couple of weeks, my mother took me back to her place in Jamaica, Queens, where we lived with my four-year-old sister, Marion, my mother's mother, my mother's other three sisters—Esther, Peggy, and Pearlie—her brother Joe, and a Swedish carpenter named Mack, who Esther later married. The whole family had come to America from Hungary when my mother was four.

My mother's father—my grandfather—was almost never referred to in that house. Rumor has it he's still in Hungary—and still drinking. My dad wasn't around much, either. I found out much later that he was a ladies' man. Dad had no time for his kids—he was always out trying to make new kids. I was born on my father's birthday. It didn't mean a fucking thing. His first wife was a southern girl. It was literally a shotgun wedding—and the marriage lasted until my father went back on the road with his vaudeville act.

I was an ugly kid. When I was born, after the doctor cut the cord, he hung himself.

My mother was my dad's second wife. She was pregnant with my older sister, Marion, so Dad did the honorable thing.

I feel awkward referring to my father as "Dad." When you hear that word, you picture a man who looks forward

TOP RIGHT: *My mother and father in one of the rare moments I saw them together.*

BOTTOM RIGHT: *As you can see I was a serious kid. I had only one thing on my mind—to play Las Vegas.*

to spending time with his family, a man who takes his son camping or to a ball game every once in a while. My father and I did none of those things. He didn't live with us. Show business kept him on the road practically all the time—or was it my mother?

When my father wasn't on the road, he'd stay in New York City. About every six months, I'd take the train from Kew Gardens into New York to see him. We'd walk around for an hour and talk—not that we ever had much to say to each other—then he'd walk me back to the subway and give me some change. I'd say, "Thank you," and then take the subway back home.

I figured out that during my entire childhood, my father saw me for two hours a year.

---

*In my life I've been through plenty. When
I was three years old, my parents got a dog. I
was jealous of the dog, so they got rid of me.*

Although I didn't realize it at the time, my childhood was rather odd. I was raised by my mother, who ran a very cold household. I never got a kiss, a hug, or a compliment. My mother wouldn't even tuck me in, and forget about kissing me good night. On my birthdays, I never got a present, a card, nothing.

I guess that's why I went into show business—to get some love. I wanted people to tell me I was good, tell me I'm okay. Let me hear the laughs, the applause. I'll take love any way I can get it.

When I was three years old, I witnessed my first act of violence. I walked into the living room and saw my mother lying on the couch, being beaten by her four sisters. My mother was kicking and screaming.

"Get Joe!" She yelled, "Get Joe!"

I did what my mother told me. I ran up two flights of stairs and started pulling on her brother Joe to wake him up. I kept repeating, "Uncle Joe, downstairs! Downstairs!" He came down and broke it up.

What a childhood I had. Once on my birthday my ol' man gave me a bat. The first day I played with it, it flew away.

From the time I was four years old, I had to make my own entertainment. There was a parking lot next to our three-story building that was always vacant after dark. Every night I would hear voices below my window, and I knew what that meant—there was going to be a fight. This is where the local tough guys would come to settle their beefs.

From my windowsill, I had the best seat in the house.

Many nights, about twenty guys would be down there, rooting for whichever fella they wanted to win. The fight itself was usually over in a few minutes—the winner would walk away happy with his pals, while the loser was left on the ground, usually bleeding, usually with a couple of his consoling buddies.

Even as a little kid I always identified with the loser. Most kids fall asleep listening to a fairy tale. I fell asleep listening to a guy yelling, "Enough! I've had enough!"

_____

I told my doctor I broke my arm in two places.
He told me to keep out of those places.

_____

My mother was coldhearted and selfish, and her sisters weren't much better. I remember being lied to by my aunt Pearlie when I was four. She was taking my sister to the movies and I wanted to go, too, but she wouldn't take me, so I pleaded and pleaded until she finally said, "Okay. Go wash your face and hands real good, and I'll take you with us."

I was so happy that I ran into the house and up two flights of stairs to the bathroom to wash my face and hands. But when I came back out, Pearlie and Marion were gone. I could see them down the block, running away from me. I stood there crying and yelling, "Pearlie, I washed my hands and face real good . . ."

8

*When I was a kid, I never went to Disneyland.
My ol' man told me Mickey Mouse died
in a cancer experiment.*

I was four years old when I got my first laugh. One night when I finished my dinner I said, "I'm still hungry."

My mother said, "You've had sufficient."

I told her, "I didn't even have *any* fish."

Most of the time, my grandmother kept an eye on me, if you'd call it that. She would be in the kitchen doing her chores while I'd be in the backyard banging nails into pieces of wood all day. Once in a while she'd glance out the window to see if I was still banging.

One day I got curious about what was on the other side of our fence, so I put my hammer down and walked out of the backyard. I walked a half block to Jamaica Avenue, the main drag in the area, and suddenly I found myself in the midst of a hustling, bustling neighborhood. I thought, *Boy, this is fun. To hell with hammering nails.*

After that, I used to walk there every day. My grandmother never noticed that I was gone.

On one of my walks—I was five by this time—a man asked me to come up to his office. After I'd climbed a couple flights of stairs, he offered me a nickel if I'd sit on his lap.

Wow, I thought, *a nickel!*

So I sat on this man's lap. He held me and then kissed me on the lips for about five minutes. Then he said, "You can go now, but don't tell anybody about this. Come by again tomorrow, and I'll give you another nickel."

I never told anyone, and I kept going back to that man every day, and I got a nickel each time. How long did this go on? I don't remember. It could have been a few days, a few weeks. Or maybe it was just a summer thing. Let's face it—at five years old, I was a male hooker.

Thanks for lookin' after me, Ma.

When I was a kid I got no respect.
When my parents got divorced
there was a custody fight over me . . .
and no one showed up.

I was ten when the Great Depression hit. Money was very tight then, so my father arranged for us to live with his mother in the East Bronx, in a really poor and rough neighborhood. She had a small one-bedroom apartment on the top floor of a six-story walk-up. My mother and sister slept in the living room, and I slept on a cot in the foyer. My father stayed at his place in New York.

School was tough.

All the kids wanted to fight and the teachers hit you, too.

My teacher, Mr. O'Connor, was a strange man. He had a beautiful voice—an Irish tenor—and when he sang "The Rose of Tralee," you loved him. It was hard to believe that this was the same man who'd tell students "you're getting one" or "you're getting two."

If you misbehaved he would call you to the front of the class. "Put your hand out," he would say, "palm up." Then he'd tell you how many times he was going to smack you with his thick ruler, depending upon what you had done and his mood.

I made sure I was a good boy that year, but I slipped just before Christmas. The whole class built a beautiful cardboard display about the yuletide season— one day, I gently touched the display, but I guess I wasn't gentle enough because my finger went through the cardboard and poked a hole in those snow-covered "mountains."

As I pulled my finger back, I could see Mr. O'Connor looking at me.

Then I heard those famous words: "All right, front of the class. Put your hand out, palm up."

I was hoping I'd just get one.

Then he said, "You're getting two."

He gave me the first one, and it hurt like hell. Then before I could recover, he hit me again.

As I was standing there in pain, my hand burning, I said to him, "How about a song?"

Then I got two more.

I like to date schoolteachers. If you do something
wrong, they make you do it over again.

Living in the Bronx, the big thrill was at night when we'd roast "mickeys." We would start a fire in the gutter against a curb, put in potatoes, and in an hour, they were delicious. We always put in extra potatoes. We knew we'd have guests for dinner. (Black tie optional.)

We lived in the Bronx for a year, then moved to a rooming house in Far Rockaway, Long Island, near the ocean, on July 19. My mother waited until that date so that she could get the place cheap—$39 for the rest of the summer. The three of us—my mother, my sister, and me—lived in one room, ten long blocks from the beach. But it was a beach.

My first day there I saw a kid I knew from the Bronx. He was selling ice cream on the beach. That became my job for the next four summers. It was hard work for a young kid—carrying around a heavy carton of ice cream loaded with dry ice so that the ice cream wouldn't melt. It was also against the law but no one cared—a minor offense—and I could make at least a dollar a day. For that kind of money, I became a criminal.

That first summer I did pretty good. I saved $100, and my mother put it in the bank for me. When I looked at my bankbook a few months later, I was shocked. All my money was gone. When I asked my mother about this, she just said that she'd needed it. And that was that. I said to myself, *Hey, what's the big deal? It's your mother.* But then I thought, *She should have at least sat me down and told me what was going on before she took it.* That would have been easier for me, but that was how my mother did things.

My old man never liked me. He gave me my allowance in traveler's checks.

After that summer, my mother wanted to live near her sister Pearlie. (You remember Pearlie. She's one of the sisters who beat up my mother.) So we moved to Kew Gardens, in Queens, which was a much nicer area. But that was a problem—it was too nice. We were much too poor for the neighborhood, and I never fit in there.

We had a one-bedroom flat. To help pay the rent, my mother took in two boarders, Max—a gangster from Detroit—and his girlfriend Helene. Max and Helene slept in the living room; I slept in the bedroom with my mother and sister.

I couldn't figure Max out. At first, I thought he was a nature lover, because he'd sit for hours just looking out the window. Then I realized he was on the lookout for trouble.

One night Max was drunk. My mother and I stood

there listening to him argue with Helene. Max was saying, "I'll go back to Detroit and be a gunman." I stood there thinking, *Boy, he's a real gangster.*

My mother had only one thought: *I'm losing a tenant.*

One time I asked Max, "Is that a real gun?"

He said, "Yeah."

I said, "You gonna shoot someone?"

He said, "Only if they ask too many questions."

I live in a tough neighborhood. They got a children's zoo. Last week, four kids escaped.

When Max and Helene left, we had to move again, to an even cheaper place in Kew Gardens, next to the Long Island Railroad station, but we were still the poor trash on a ritzy block. Not being able to keep up with the other kids financially made me feel inferior. The kids I went to school with would see me delivering groceries to the back door of their homes, so they looked down on me. I couldn't play football in school because I didn't have money for the equipment. Tennis was out. That cost money, too, so I played a game called "stoop." You throw a ball against the stoop and try to make it land where your opponent isn't. No investment required—except for the ball—the stoop is always there, always free.

*A homeless guy came up to me on the street,
said he hadn't eaten in four days. I told him,
"Man, I wish I had your willpower."*

Growing up, I got no guidance from my mother. The only advice I can remember getting from her was, "Never eat a frankfurter from the man on the corner with the orange umbrella. Those hot dogs are made of snakes."

I believed her. I was a kid. What did I know?

My father's place in New York was a half-hour subway ride from Kew Gardens. Once when I was about eleven, I went into town to see him. We were sitting at the counter in a drugstore—I was drinking a Coca-Cola; he had nothing. When I was done, he went outside while I paid the check. While I was looking for the waitress, I saw that I could walk out without paying, so I did. When I got outside, I was pretty proud of myself, and said to my father, "Hey, I beat the check."

My mother would have congratulated me, but my father wasn't going to. Disgusted, he said, "Who taught you that, your mother?"

He made me go back and pay for my soda.

I tell ya, I grew up in a tough neighborhood.
The other night a guy pulled a knife on me.
I could see it wasn't a real professional job.
There was butter on it.

A couple of months later, my mother and her sister Esther took me on a trip to Atlantic City. (You remember Esther. She's another one of the sisters who beat up my mother.) But this wasn't going to be a nice day at the beach with the family—it was an ambush. They had surmised that my father would be there on the beach with his girlfriend Lily. And they were right.

We got to the beach, and sure enough, there was my father with the big love of his life. My mother and my aunt went over to them and started yelling, "Look at him! He's a tramp!"

As you might expect, a crowd gathered to watch this free freak show. "*This* is his wife!" my aunt screeched, pointing at my mother. And then she pointed at me. "He left this ten-year-old boy to be with that whore!"

With that, the crowd looked at me *like it was my fault.* It was awful. I wanted to crawl under the boardwalk. I was standing there sweating, sand in my shoes. My aunt said to me, "Jackie, when you grow up, punch him in the nose."

What a day that was. But I tried to look at the bright

side—for a few minutes I got to see my father. I knew I wouldn't be seeing him again for six months.

I was an ugly kid. I worked in a pet store.
People kept asking how big I get.

I was always hustling to make some money because my mother never gave me any. I started delivering groceries when I was ten. The grocer gave me a choice—ten cents an hour or three cents for every order I delivered. I worked fast, so I took the three-cents-an-order deal. But I soon realized that I'd made the wrong decision. Every order he gave me was at least ten blocks from the store. Two hours later, I told the guy, "No more three cents an order. From now on, I'll take the ten-cents-an-hour deal." He said okay.

Then he gave me six deliveries—all going to the same apartment house.

When I finished for the day and was ready to go home, the boss turned out to be a good guy. He gave me an extra buck and a Swiss-cheese sandwich for the walk home. The sandwich he made tasted real good. He put something on it that I'd never had before—lettuce.

I tell ya, my wife's a lousy cook.
After dinner, I don't brush my teeth.
I count them.

My entire childhood, my mother never made me breakfast. She would sleep every day until about eleven. She never got up to see me off to school, so before getting on the bus in the morning, I'd go to a place called Nedick's, a big cafeteria chain at the time. I had the same breakfast every day—orangeade, doughnut, and coffee—for ten cents. For me, this was a home-cooked meal.

Before I went to school in the morning, I would take care of a newsstand. I would make change for people who bought papers or cigarettes. Sometimes, my fingers were so frozen that I'd ask the people to make their own change. For this, I got a dollar a week and a nickel piece of candy.

As a kid I felt inferior to everyone. I was too shy and insecure to talk to girls. At around fifteen, there was a girl I liked, but the thought of speaking to her was petrifying. I would see her on the bus, and she really turned me on. I always made sure my books were on my lap.

I couldn't talk about sex with my mother, and my father had his own life in New York, which didn't include me. (As I grew older, I realized that he was a smart guy.) When vaudeville died in the late thirties, my father became what they called a customer's man in the stock market. He was basically a stockbroker—he invested money for a very elite clientele. He did extremely well for many years.

My mother would never listen to my problems and had no interest in how I was doing. I remember giving her my report card to sign one time. My marks were pretty good, and I was looking forward to some praise, but she didn't give a damn. She just signed the report card without a word and gave it back to me.

I said, "Don't you want to look at it?"

She said, "You know what you have to do."

What a childhood I had.
My mother never breast-fed me.
She told me she liked me as a friend.

When you're a kid, you have no way to compare mothers. I knew only mine, and to me she was the best. My entire life she never gave me a gift. But I remember I gave her one.

When I was fourteen I tried to start a newspaper and I failed, but I did have an egg route. Although I didn't make much money, I put aside enough to buy my mother something for Mother's Day. She liked to sit in the kitchen every night and have a few beers. So I bought her a six-pack, and I made my own card:

Dear Ma,
Happy Mother's Day.
Let's hope I keep selling egg after egg,
So you can keep drinking keg after keg.

Okay. I'll go back to prose.

What a childhood I had. My parents
sent me to a child psychiatrist.
The kid didn't help me at all.

My aunt Pearlie and her husband had a drugstore–soda fountain in Astoria, Long Island. After they'd worked that for a while, they leased the bar and grill next door. When my mother would visit Pearlie, they'd hang out in the bar.

I was around fourteen or fifteen at the time, and I sometimes worked behind the soda fountain at Pearlie's drugstore. On my break one night, I walked down the sidewalk and looked through the window of the bar and grill. There I saw my mother sitting in a booth, having a drink with some old guy. What I saw next was even harder to believe. When my mother's drinking companion wasn't looking, she'd throw her drink under the table. A few moments later she'd ask for another one.

That was my apple-cheeked mom—hustling drunks for her sister.

I tell ya, my family were always big drinkers. When I was a kid, I was missing. They put my picture on a bottle of Scotch.

Both my father and his brother were in show business. They did a pantomime act in vaudeville, one that involved breaking a lot of plates, as I recall. When Eddie Cantor, the big star of yesteryear, was seventeen, he was broke, hungry, and trying to get into show business. My uncle Bunk took a liking to the kid, and took Cantor on the road with him.

One night, a singer on the show had one too many and couldn't go on, so my uncle Bunk said, "Put Eddie on. He can do a couple of songs."

So Eddie went on, did his best number—"Susie"—and the house went crazy. They wouldn't let him off the stage. From that night on he was "Eddie Cantor," and nothing could stop him.

Cantor never forgot what my uncle had done for him,

and Uncle Bunk was on Cantor's payroll for the rest of his life.

When I was fourteen, my uncle Bunk arranged for me to be in the audience for the taping of *The Eddie Cantor Show*, a half-hour variety show they shot in New York City. Every week I got the best seat in the house—front row, center.

But one night my uncle Bunk said to me, "Jackie, we got a problem. You can't sit in the front row anymore. After the taping last night, Cantor said to me, 'Who is the kid who sits in the front row every week and never laughs?' I told him you are Phil's kid. To make it short, Jackie, you can't sit in the front row anymore. You have to sit in the back."

The next week, when I showed up for the taping of Cantor's show, uncle Bunk put me in the last row of the theater. When the show started, no matter what Cantor said, I just kept laughing and laughing.

My uncle came over and said to me, "Why are you laughing so much?"

I told him, "I wanna get my seat back in the front row."

My ol' man took me to a freak show.
They said, "Get the kid outta here.
He's distracting from the show."

When I was about fifteen I got a job as a barker for a theater, the Academy of Music, on Fourteenth Street in New York. I was the fellow standing outside yelling, "Just in time! Feature's going on in ten minutes! Plenty of good seats! Win a twelve-piece set of dishes!"

What made it really tough was that the boss had his office on the fifth floor, and if he didn't hear me, he'd call down to the assistant manager, who would tell me to bark louder. I was yelling all night so that the boss could hear me five flights up—and he had his windows closed because of the cold.

When I was sixteen, I was taking a walk down Broadway in New York City and I saw a sign—10 CENTS A DANCE. I didn't know what exactly went on in those places, but I did know that they were for adults, but I thought, *I'm big for my age. I'll give it a shot, see what it is.*

The woman in charge met me at the door. She was extremely nice and walked me over to the girls. Right away, I saw one girl I really liked, so the two of us sat at a table and started talking. She seemed to really like me, too. She even held my hand. I thought, *Hey, I'm doing all right!* I was in love. We sat there talking for about ten minutes like a couple of lovebirds.

The next thing I knew, the nice manager came over to our table and said, "We have to get paid now. Your bill so far is six dollars."

I said, "Six dollars? For what? I didn't dance."

I turned to the girl I'd fallen in love with and said,

"What's going on? I didn't dance, and I don't have any money."

Now the nice woman in charge and the girl I was in love with started yelling at me. The girl of my dreams screeched, "You fuckin' idiot, you cost me money! I could have been dancing with someone else!"

Then the manager yelled, "Punchy! Get this creep outta here."

I tell ya, the first time you see Punchy, you hope it's a bad dream. But as he hustled me out of there, I tried to be funny. I said, "Maybe I'll open my own place—eight cents a dance. When you come to my place, you'll see big Hollywood stars."

Punchy said, "Kid, you keep talking, *you'll* see plenty of stars."

With my wife, I got no sex life. She cut me down to once a month. Hey, I'm lucky—two guys I know she cut out completely.

At eighteen, I got my driver's license, and I bought an old Ford with a rumble seat. They don't make them like that anymore, thank God.

That car gave me nothing but trouble, but one thing I'll say, it was very easy for me to find it. It was always on a lift. I was always watching it, going up and down, up and down. I had the only car that had more miles on it vertically than horizontally.

When my wife drives, there's always trouble. The other day she took the car. She came home. She told me, "There's water in the carburetor." I asked her, "Where's the car?" She said, "In a lake."

Sometimes my car became a taxi. I noticed that the Long Island Railroad station had very few cabs, so when the trains would pull in and there were no cabs, I'd offer to drive people home. I'd tell them, "Pay me whatever you think is fair." Most of them knew exactly what a cab would cost, and that's what they'd pay me.

I got a couple of other jobs, too. Monday and Saturday I drove a laundry truck. Monday was pickup day, Saturday was delivery. Thursday and Friday, I drove a fish truck for the Little Fish Market in Kew Gardens.

One afternoon I was delivering some fish orders to Jamaica Estates, a rich neighborhood in Queens. I was stopped at a red light when a classy, good-looking chick in a sharp car pulled up alongside me. She smiled at me and waved, then signaled for me to follow her.

I thought, *Oh, man, did I get lucky! I should have taken a shower!*

I followed her until she pulled into the driveway of a big, expensive house. I parked my truck behind her car, and snuck a peek in the rearview mirror to check my hair. Then I stepped out of the truck and walked toward her.

"Hi, honey," I said. "Where do we go from here?"

She said, "Nowhere. I just thought I'd make it easy for you to find my house. You have my fish in your truck."

I was gonna say, *"Honey, you should see what I've got in my pants."*

All my wife and I do is fight about sex.
The other night, we really had it out.
Well, I'll put it this way—I had it out.

RIGHT: This is me sixty years ago. As you can see, I haven't changed at all.

Chapter **Two**
How Can I Get
a Job Like That?

I tell ya, I don't get no respect. One time I was workin' a club, and the manager said he'd pay me under the table. I waited down there for two hours. He never showed up.

When young people ask me, "How do I become a comedian?" I have to tell 'em: it's not easy. Everyone struggles when they are starting out, even Andy Kauffman, Sam Kinison, and Jim Carrey. Jim Carrey, who's about the most talented guy I've ever seen onstage, opened for me for two years on the road, and I remember plenty of nights when he couldn't get a laugh. But he didn't give up.

To be a comedian, you have to get on the stage and find out if you're funny. The thing is, how do you get on the stage? Here's one way: Get a job in a local comedy club as a waiter—or anything. Observe the comedians as much as you can—even down to studying the way they walk on- and offstage. See how they approach their material and what their attitude is. What will your attitude be when you walk up to the mike? Should it be joyous? Should it be troubled? Figure out what attitude fits you best.

---

*In high school, when I played football I got
no respect. I shared a locker with a mop.*

The next thing you gotta do is start writing your own stuff. If you can't write your own material, you have very little chance of making it as a comedian. When you're starting out, just try to get five minutes of good material, then work on it and work on it until you think it's great. By now you should be friends with the owner or whoever runs the club you've been hanging out in. When you feel you're ready, ask if you can go up and do five minutes. If he likes you, he'll put you on. He may not pay you, but he'll put you on. It may be at 3 A.M. on Monday morning, but he'll put you on. And now it's up to you to show how funny you are.

From the moment you walk onstage, try to make the people like you. That's the most important thing. If they like you, you can get a big laugh with a mediocre joke. If they don't like you, you've got some serious thinking to do about your career choice.

That's how you become a comedian. It's the hard way, but it's the only way. As Peter O'Toole said in the movie *My Favorite Year,* "Dying is easy. Comedy is hard."

I began writing jokes when I was fifteen. I think I was so unhappy all the time that I was trying to forget reality with jokes. I was always depressed, but I could tell a joke and get a laugh. But not from my mother. She never thought my jokes were funny. I'd write a joke and tell it to her. Nothing. I'd never get a laugh.

I can remember only one time that my mother had a good laugh after something I'd said. I told her that one of my friends had told me, "Hey, Cohen, don't let 'em shit all over you. Open your mouth."

That made her laugh.

When I was a kid, I got no respect.
I told my mother I'm gonna run away
from home. She said, "On your mark . . ."

A round that time, I started performing in amateur shows. In those days I did a Chinese act—On Too Long. My act consisted of two impressions: W. C. Fields and Al Jolson. My stage name then was Jack Roy.

I got my first paying job as a comic when I was eighteen.

The agent's name was Jack Miller. Jack would pack ten acts into his car and drive us from New York City to a theater in Newark. Everybody got paid the same—two dollars—and Miller paid all his acts in quarters. Why quarters? No one ever knew.

I remember my opening joke that first night. I said, "Would you look at the audience we got here tonight? All these women, they look like a beautiful bed of roses. Of course, there's a weed here and there."

I guess I did okay that night—I got all eight quarters.

At nineteen, I landed my first job in the Catskills, a big resort area north of New York City. I worked ten weeks at $12 a week, plus room and board. My act was a lot of jokes, some impressions, and a few songs.

That gig lasted through the summer, and after that I couldn't get booked anywhere. All day I walked around in the heat, going from agent to agent, trying to get a job in show business. After three weeks I gave up. I had to make some money. It was back to the laundry truck and fish truck.

But when I wasn't behind the wheel, I was writing jokes and trying to get in good with the small-time agents, who could book me on a Saturday night for five dollars— less their 50 cents commission.

It was rough. Even then, with hundreds of clubs in New York, no one would take a chance on a kid starting out. And the few times I was lucky enough to get a job, the places I worked in were tough. How tough? If you didn't do good, they would pick a fight with you. I remember one bouncer said to me, "You're an asshole."

I said to him, "If I'm an asshole, there's a reason for it . . . you're contagious."

When I woke up, the first face I saw was my dentist's.

When I was a kid I worked tough places—places like Fonzo's Knuckle Room, Aldo's, formerly Vito's, formerly Nunzio's. That was a tough one, Nunzio's. I sat down to eat. On the menu, they had broken leg of lamb.

One time I was booked in a place in the Bronx called the Neck Inn. I was supposed to get five dollars for emceeing two shows on Saturday night. When I arrived that night, the boss greeted me with the news that he didn't need me because one of his waiters was going to emcee the shows instead. He was nice enough to give me ten cents for my trip up to the Bronx and back—the subway was a nickel back then.

I was very depressed because I had been counting on the $4.50. Being a true thespian, though, I asked the boss if I could be a waiter for the night. He said okay, and gave me four tables in the back. I made $2.50 that night. Not bad, but some of the waiters really made out. There was a two-piece band in the club—piano and drums. Some of the waiters sang with the band, and people threw money at them.

I could sing, so I thought, *How can I get a job like that?*

On Monday, I went to the Entertaining Waiters' Union in Manhattan and pleaded with the man in charge for a singing waiter's job. His name was Ed Delaney, and he was cold to me. His union was practically all Irish, and you might say that I didn't fit in.

In my desperation, I pulled three dollars out of my pocket—all the money I had—and said, "Mr. Delaney, look. This is all I got." I put the money in his Christmas Fund jar. "I really need this job bad."

He softened up. He said, "Okay, kid, let me hear you sing."

I sang two songs. I must have sounded okay, because Delaney gave me a job.

That was my first big break in show business—a job as a singing waiter at the Polish Falcon nightclub in Brooklyn. I worked there Friday, Saturday, and Sundays, earning $20 or $30 for the weekend. In between songs, I would try out jokes.

I also worked in several other places as a singing waiter just so I could break in jokes. One time, it was a Saturday night, and the place was packed. I told a couple of jokes, the audience laughed, so I just kept telling jokes.

As I walked off after two songs and about ten minutes of jokes, the boss was waiting for me, and he was livid. "What the fuck are you doing?" he said. "You're a waiter! You want to tell jokes, tell 'em to your fuckin' friends! These people are thirsty! Get 'em their fuckin' drinks!"

With my wife I don't get no respect.
I made a toast on her birthday to
"the best woman a man ever had."
The waiter joined me.

One of the strangest nightclub owners back then was a guy named Meyer Horowitz. He owned a club in Greenwich Village called the Village Barn. Business was so bad there that he had to let his chef and two waiters go. But his customers never knew it. When a couple would walk in, Meyer would be working behind the counter. They'd sit down, he'd take their order, write it down on the check, then walk over to the kitchen door, open it, and yell out the order into the back—to no one. Then Meyer would come back to the counter and do some work. Three minutes later, he'd go back to the kitchen door and yell, "I'm waitin' on the roast beef!" Then he'd go into the kitchen. A few minutes later he'd come out with the sandwich—which he had made himself.

When Meyer's place got busy, watching him work was a thing of beauty. He'd be going back and forth, back and forth—a cup of coffee here, yelling into the kitchen here, a sandwich there, a piece of pie over here, back to the kitchen, a cup of soup over there . . .

---

*My wife told me she likes to have sex
in the backseat of the car. I drove
her and that guy around all night.*

To get work in those days, I had to spend a lot of time in and around Boston. There were plenty of nightclubs up there, and I could work steady at $100 a week, less 10 percent commission, travel expenses, and a place to stay. But hey, I wasn't complaining. I was in show business, I was young, and I had the optimism of youth.

After doing that for a few years, though, I decided it was time to hit New York City again and go for the big time. This meant walking around all day, seeing all kinds of small-time agents, trying to get booked for the weekend, or even just a Saturday night.

No luck.

Once again, I was driving the laundry truck on Mondays and Saturdays, and the fish truck on Thursdays and Fridays. I had Tuesday and Wednesday off, so that's when I'd go into New York and hang out in front of the 46th Street Pharmacy in Times Square. That's where all the comedians were, and I tried to learn as much as I could from them.

My uncle's dying wish,
he wanted me on his lap.
He was in the electric chair.

After about two years, I had improved my act to the point where I could work pretty steadily, earning approximately $150 a week—more when I worked out of town.

I worked out of town a lot, and whenever I'd come back to New York, other comics would ask me, "How was the job?" They didn't want to know about the club, or the crowds, or the pay. Comedians judged a job by how well they did with the girls. If a comic was out of town for a week and didn't score, he'd say, "What a joint! Forget it. Don't go up there. It's nothin'."

But if the comic did a job out of town and slept with three or four chicks during the week, he'd say, "Man, you gotta play this place! It's great!"

I can't explain it, but there's something about the aura of show business that really helps with the girls. If a guy is only fair looking, he becomes good looking when he goes into show business. Not me, but most guys. Girls treat me like I'm their father—they keep asking me for money.

I told my doctor I want to get a vasectomy.
He said with a face like mine, I don't need one.

One of my best memories of working out of town was of a nightclub in Chicago called the Silver Frolics. After my first show, I joined the boss for a drink at the bar. There were about forty people at the bar—mostly girls, and most of them not hookers. It was a place where people came to hang out, and who knows who you'd meet?

As we were drinking at the bar, the boss said, "Kid, I like you. If you see a girl you like, let me know." I immediately put down my drink and started checking out the crowd. I was walking around looking at legs, arms, shoulders, boobs . . .

When I saw someone I liked, I said to the boss, "Who's the girl over there?"

He yelled out, "Viola!" And she came right over.

Viola and I went back to my room and frolicked our brains out.

I tell ya, with my wife, I got no sex life.
Her favorite position is facing Bloomingdale's.

I was working some nightclub in Asbury Park, New Jersey, when I was around twenty-one. After the show, I saw this girl. Our eyes met and it was clear that we were both thinking the same thing: *Now!*

But where were we gonna go? Neither of us had a private room, so we drove down to the beach. We made ourselves comfortable on the sand and started going at it . . .

I don't want to go into too many details, but I met the nicest cop in the world that night. He let me finish. Then he shined a huge flashlight in my face.

So what did I do? I sang "Mammy"!

No, I didn't. But I should have.

That same night, I also found out that I had the nicest booking agent in show business. I had my bed, he had his. The girl I had gone to the beach with still felt the same way I did—we wanted more. So we went back to my room. My agent was already in bed, so I quietly opened the door and peeked in. I could see that he was sleeping, so I told the girl, "Come in. It's okay. Just be quiet."

We slipped into my bed and started warming each other up. Everything was going great. I was ready for seconds, and then she told me to put on a rubber. "I don't have any more rubbers," I whispered. "I used my last one on the beach."

She said, "If there's no rubber, forget it."

Just then, my agent gets out of bed, goes to the closet, takes a rubber out of his coat pocket, hands it to me, and goes right back to bed without a word. Now, that's a booking agent.

I asked my wife, "Last night, were you faking it?"
She said, "No, I was really sleeping."

Sometime later I was working in New York, at the 78th Street Tap Room. It was a very small club—about a hundred people—but business was good there. One night after I finished my show, the waiter came over to me and said, "Jack, there's a man over here with a girl. He'd like you to sit down and have a drink with them."

Next thing I knew, I was having a drink with a young girl—about twenty-five—and an old man, about seventy. We were sitting there talking for a few minutes when the guy said, "Look, here's the situation: I'm too old to do anything sexually, but my girl would like to. I only live two blocks from here. When your show is over, come back to my place. You can be with the girl—she likes you—and you'll both have a good time."

I was thinking, *What kind of a weird thing are they into?* But the guy looked harmless, and she looked hot, so I decided to go for it.

After my show, we all went back to his apartment, and the girl and I went into the bedroom. I closed the door, and we got undressed. But before we got into bed, I wanted to make sure I wasn't being set up, so I opened the door and took a peek outside. The old guy was just sitting in the living room, reading the paper.

I closed the door and wow, did I get lucky! From then on, that girl and I didn't say another word—our tongues were too busy.

Thank you, show business.

By the way, it may seem like I'm getting a lot of girls, but remember, this is over a period of sixty years.

They took a survey: "Why do men get up in the middle of the night?" Ten percent get up to go to the bathroom and 90 percent get up to go home.

Chapter **Three**
Plans for
Conquering the World

*This girl was fat. I hit her with my car.
She asked me, "Why didn't you go
around me?" I told her, "I didn't have
enough gas." I mean fat. She was
standing alone. A cop told her to break
it up. She stepped on a scale, a card
came out. It said* One at a time.

When I was a kid, comedians were real characters. I remember a most unusual "gentleman" named B. S. Pully. I'd heard plenty
about B. S. Pully before I met him. People
said he was a low-class, filthy, dirty, funny maniac. When
I met Pully, I learned that they had all been too nice.

I was eighteen, hanging around New York at night trying to learn about show business. I can't remember how it
happened, but I wound up in an amateur contest in a
nightclub on Fifty-second Street. B. S. Pully was the master of ceremonies.

I entered the contest as a singer.

When it was my turn to go on, Pully said, "Our next contestant is gonna sing for us. Give a hand to Jack Roy."

I walked onto the stage and stood next to Pully, who, in a voice that sounded like someone shoveling gravel, said, "What song you gonna sing, kid?"

I said, "'You Are Always in My Heart.'"

Pully said, "All right, kid. I'll be in your ass later."

Years later I was working in a New York nightclub

Hey, I'll take it any way I can get it. Just ask my wife.

called the Living Room. It was a popular place, and all the acts liked to work there, but it was a very small room, and everyone could hear everything. In the middle of my act one night, a phone rang in the audience.

Ring . . .

"Hello," said Pully in his gravel voice.

Everybody in the audience turned to listen to Pully, so I just stopped my act and stood there.

Then we all heard Pully say, "I told you not to call me here, you rat bastard! The show is on!" Then he hung up the phone and said, "Go ahead, kid."

One night Pully was at a fancy social function. How he got in I'll never know. Anyway, he asked one of the prominent society women to dance. She accepted, not knowing that Pully had taped a Coca-Cola bottle to the inside of his right thigh. As he was dancing, he would look straight into the woman's eyes and dip so that the woman's thigh would rub up against the bottle.

For the rest of the dance, Pully just kept staring at her, cool as can be, occasionally doing his famous dip. Years later, B. S. Pully appeared in the movie *Guys and Dolls*. He played Big Julie.

When Pully did his act in a nightclub, another gentleman often joined him. This guy called himself H. S. Gump. That's right—Bull Shit Pully and Horse Shit Gump.

After the nightclubs would shut down for the night, many of the acts would hang out at Kellogg's Cafeteria on Forty-ninth Street. One night we were all sitting at a table having an early breakfast and Gump, who was drunk, said loudly, "Where's the salt?"

My friend Martin handed Gump the salt and said, "You want the pepper, too?"

Gump said, "Fuck the pepper."

"If you fuck the pepper," Martin said, "your cock will sneeze."

Martin had a strange sense of humor. His full name was Martin Nadell. He invented Jumble, the scrambled-word game, which has nothing to do with this next story. I was working in a nightclub in the Bronx called the Red Mill. Opening night, in the middle of my act, the next act—a stripper who worked with fire—came walking through the audience, heading backstage, carrying her lit torches. The audience saw the girl with the fire, and forgot all about me. You might say it was distracting.

After the show, I knocked on her dressing-room door. When she opened the door, I asked her if she'd wait until she was backstage to light her torches.

She got very huffy. "Don't tell me what to do!" she said. "I fuck you! I fuck everybody!"

A short while later, there was another knock on her door. She opened it, and Martin was standing there naked.

She said, "What the hell is this?"

Martin said, "You said you fuck everybody, so I figured I'd be first."

I tell ya, I got no sex life. My dog watches me in the bedroom. He wants to learn how to beg. He also taught my wife how to roll over and play dead.

In the forties and fifties, Hansen's Drugstore at Fifty-first Street and Broadway in New York was where every kind of performer hung out during the day—actors, actresses, comedians, tightrope walkers, whatever you wanted. They were all there in the afternoon, talking show business and perfecting their plans for conquering the world.

There were many colorful characters there, but we all agreed that the most colorful was a guy called Tootsie.

He got that name because he was always singing an impression of Al Jolson. He would sing, "Toot-Toot-Tootsie, good-bye, Toot-Toot-Tootsie, don't cry . . ."

Tootsie told everybody he was a big, big agent. He would sit down at your table, open a large portfolio, and show you pictures of his big clients. The first one was a publicity shot of Van Johnson. Tootsie would say, "Van Johnson. Nice boy to have in your stable, right?" Then he'd turn the page. "Who's this? Ginger Rogers. Good girl to have under contract. We're very close, you know, very close." Next would be Clark Gable. He'd say, "Oh, what a guy. We've been together over thirty years."

And he'd continue to roll out these pictures of the biggest stars of the day and say things like, "I'm getting her a three-picture deal at Paramount . . ." or, "He's going to headline in London for a month . . ."

Often he would walk up to a comic and say, "Are you available on September twenty-fourth for two weeks?"

The guy'd say, "Yeah."

"Okay," Tootsie'd say, "I'll get back to you. I think I got something good for you."

Then he'd turn to the next fellow and say, "Are you open October first for a weekend in Pittsburgh?"

"Uh-huh."

"Good. I'll get back to you . . ."

He never got back to anyone. But that was okay—everyone knew that he was out of his mind.

Next door to Hansen's Drugstore was a small restaurant called B&G. They had handmade signs advertising their food Scotch-taped all over their windows. In that neighborhood, just a couple of blocks north of Times Square, there were always a lot of out-of-towners walking around, taking in the sights of New York. They'd stop and look at the signs, and if they liked what they saw, maybe they'd come in to eat.

To have a few laughs, we'd make up our own signs and tape them over the real ones. Ours would say, BEST FUCKIN' HAMBURGER IN TOWN! or OUR SOUP WILL KNOCK YOU ON YOUR ASS! Then we'd stand on the corner and watch the tourists' reactions.

That was our excitement for the day. That's what you do when you can't get a job in show business.

I was an ugly kid.
My mother breast-fed me
through a straw.

Chapter **Four**
Very Naked
from the Waist Up

My wife and I, our relationship
is on and off. Every time I get
on, she tells me to get off.

When you're starting out in show business, you go through many frustrating experiences. Today I can think back and laugh, but at the time, it was serious business.

I was working a club on Long Island once. The show consisted of me and three gay guys who had a dance act. The lead dancer was named Paris, and he had talent. In fact, he was the whole act—the other two just hung on for the ride.

After the second night's show, the boss put his arm around me and said, "Jack, I like you very much, you son of a gun. I want you to come back next weekend. You'll work Friday and Saturday again."

"Thanks," I said. "You just made a comic happy."

He said, "Come with me. I wanna talk to the dance act."

We walked over to the dancers. The boss still had his arm over my shoulder, still telling me how great I was. When we got to the dancers, he said to Paris, "I'm bringing Jack back next Friday and Saturday." He smiled at me and said, "I love you, you son of a gun." Then he told Paris, "I'd like you to work with Jack, but I can't use the other two guys in your act. I want you to work alone."

Paris stood up and said, "We're an act. We don't break up."

The boss said, "You don't break up, huh?" He then took his hand off my shoulder and said, "Jack, you're out."

I had a date with an inflatable girl.
Now I got an inflatable guy looking for me.

When I was twenty-one, I worked at a nightclub in New Bedford, Massachusetts. The show consisted of me as the comic emcee and a stripper named Virginia Kinn. After the first show there was a knock on my dressing-room door. In walks Virginia, very upset and very naked from the waist up. She said the boss had told her to cover her nipples during her act because he was worried about the vice squad.

She told me this was a huge problem for her because she had four high-rolling friends coming for the second show. "They're driving an hour and a half to get here," she said, "and if they don't get to see my nipples, they'll be very disappointed."

It was hard for me to believe what I was hearing. And due to her attire, it was hard for me to really *hear* what she was saying, so I just stood there mumbling things like, "I'm sorry . . . Oh, really? . . . Uh . . . You were saying . . . ?"

The whole time she was talking, I never thought of coming on to her. I figured, what would she want with me? She had guys with big money driving a hundred miles just to see her nipples.

As it turned out, everything was okay. Her friends started drinking, and never missed her nipples. Virginia was relieved, but I wasn't.

She was a wild girl. I took her to a bar.
She gave the mechanical bull her phone number.

In those days, I never knew when or where sex would pop up. One night, I was on the subway going home after a show. It was late, about three in the morning, but when I got off at my stop, I noticed that a girl was following me. I wasn't afraid of some little girl, so I went up to her, and we started talking. After a few minutes of conversation, it was clear that she wanted us to get together, so we went to an isolated place behind a monument in the park nearby and did it. When we were done, we went our separate ways.

On the way home, I suddenly became worried, because I didn't use a rubber, so I went to a hospital across the street. I asked to see a doctor, and a woman in a white coat came out. I told her I'd rather see a male doctor, but she assured me that she was a real doctor and that I had no reason to be concerned because she was a woman.

Well, I thought I had *plenty* of reason to be concerned, but I had no choice, so I told her about my "one-monument stand" earlier that night and said that I was worried about catching something. She had me take my pants off, and she looked at my penis, then took a hold of it to examine it more closely. At that point I started to get excited, which she pretended not to notice.

After this extremely intimate examination, she told me I could put my pants back on. "Nothing to worry about," she said. "You're okay."

I said, "So are you. What time do you start work tomorrow?"

I said to a girl I'd been seeing, "Come home with me, honey, and I'll show you where it's at." She said, "You'd better, because the last time I couldn't find it."

RIGHT: People talk about safe sex. To me, safe sex is when all the car doors are locked and her husband is dead.

During those days, I kept a small blue bulb in my glove compartment. In case sex did pop up, I had the right lighting. One night, it popped up in Baltimore. I was working at the Club Charles there when I got lucky with one of the waitresses. We decided to go to her place when we were both through with work.

When we walked into her apartment, I sat down on a chair in her bedroom. She said, "Excuse me," and went into the bathroom.

I sat there waiting.

Ten minutes went by.

I thought she was taking kind of long.

I'm sitting there.

Now it's fifteen minutes.

I thought, *What's going on?*

It became twenty minutes. Finally I got up and knocked on the bathroom door. I said, "Are you all right?"

She said, "Oh, you can come in."

I opened the door, and she's washing her stockings.

I thought to myself, *How urgent is her passion for me?*

I felt like I was one of her chores for the night. *I'll do my nails, do my hair, wash my stockings, bang him, and go to sleep.*

I was twenty-two, working at a nightclub in Bridgeport, Connecticut. That was a long ride from New York, but I wasn't complaining—in addition to my meager salary, I got a free room above the club. To my delight, also living in a room above the club was an easy-to-look-at waitress.

It is now five days later. I have been turned down by this waitress at least ten times, and the job will be over in two days. That was bad enough, but what really hurt was that she had made it with almost every other guy at the club—both bartenders, the boss, the boss's son, even the dishwasher.

I was determined to get this girl. I felt like I was a big-game hunter; I was the predator and she was the prey. Sitting in my room that night, prompted by the heat of youth, I devised a plan: I'll knock on her door, and if she lets me in, I'll charm her, and when the timing is right, I'll make my move.

I walked down the hall and knocked on her door. *This is it,* I told myself. *Showtime.*

She said, "Who is it?"

"It's Jack," I said, "the emcee. I want to talk to you."

"Nah, I'm sleeping."

"I have to talk to you," I said. "Open the door. It's important." Important to me, not her.

She opened up, and as I walked in, I told her, "I'll sit in that chair, and you can stay in bed." I figured I had a head start if one of us was already in bed.

She sat on the edge of her bed and lit a cigarette as I started talking. I figured I had at least five minutes to warm her up while she smoked, so I was flattering her as much as I could. I never knew I was such a great liar, and after a while I made my move.

She said, "What if I scream?"

I said, "If you scream, that won't help you. Everyone will just think I'm a great lover."

She wasn't impressed. She said, "I'm looking for a man who will love only me, someone sensitive, romantic, handsome, and considerate."

I said, "If I had all those qualities, I wouldn't be with you."

I tell ya one thing—I know how to satisfy my wife in bed. I leave.

In the forties and fifties, before television became big and killed nightclubs, New York was a wild place to be a performer. In the five boroughs of New York and in New Jersey, there were about three hundred nightclubs—and they all had shows. On Fifty-second Street in Manhattan, there were eight nightclubs on that one block alone.

These nightclubs used all kinds of acts, from knife throwers to fire-eaters to a girl named Rosita, who worked with a snake. Every night when the show was over, the boss would say, "Rosita, we're closing. Tell the snake to wrap it up."

I worked at a nightclub in New York called the House

LEFT: I am holding a ferocious lion that I captured in Africa. His name is Rodney Jr., and right now he resides at the MGM Grand in Las Vegas.

of Scheib's. Every Tuesday night, they had a mambo contest. One of the girls there was an excellent dancer, and very sexy, so I made a point of getting to know her pretty well. We'd have a drink or two after the show, and we got along great. After a couple of weeks, I said, "Let's get together." She said, "Sure," and then I *really* got to know her.

Before long, we were seeing each other every Tuesday after the show. Then I started seeing her on other nights, too. We were both having a real good time together, but one night she said, "I'll be getting married soon."

I said, "Good luck. Who are you marrying?"

"You wouldn't know him," she said. "He's a cop from Long Island."

I knew only one cop from Long Island. I'll call him Pete Hartmann. We were kids together—we would go to the beach, lie on a couple of comfortable rocks, and repeat all the funny lines from our favorite movies, especially the Marx Brothers. We'd laugh for hours. I don't know why, but I asked the girl, "What's his name?"

She said, "Pete Hartmann."

I was in shock, and I felt awful for Pete. I asked myself, *Should I tell him? I knew the guy years ago. Maybe he's deeply in love with this girl. He could very easily say, "She's a good girl. What did you do, twist her mind around?"*

I was also thinking, *He's a cop. He's got a gun. Who knows how he'll react?* But I did what I had to do.

The next day, I called him and said, "Pete, I have to talk with you."

We met for a drink, and I told him all about "his" girl.

He sat there quietly. When I was finished, he didn't say a word for about five minutes. Then he got up, shook my hand, and said, "Thanks." That was it. I didn't hear from him again for years and years.

Many years later, Pete and I reconnected. He told me that he'd married a different girl—not the dancer—and was doing okay, kids and everything. He told me his wife was lovely, and said, "Do you want to see her picture?"

"Oh, no," I said. "I'm not going through that again."

I tell ya, my wife likes to talk during sex.
Last night, she called me from a motel.

I spent the next few years working and improving my act and trying to get a better reputation as a comic. I wasn't always successful at that last part. When I was twenty-five, an agent named Billy Goldes booked me into a place in Montreal called the Esquire Club.

I go up there, do the first night's show, and I die. Nothing. And then I realized that no one spoke English. They all spoke French.

The next night, same thing. I died both shows.

After three days of this, I called Billy Goldes. "What did you book me up here for?" I said. "No one speaks English!"

"Yeah, I know," Billy said. "I don't like the guy who runs that club. He did me wrong on something, so I booked you up there to get even with him. I knew you were gonna die."

My agent was booking me to get even with people. That gives you some idea of how my career was going.

I tell ya, comedy is in my blood.
I wish it was in my act.

But I stayed at it, taking any job I could get. Eventually, I was on the road all over the country. I remember one gig in Shreveport, Louisiana. I arrived the night before and had a drink at the club with the boss. I had heard that this guy owned half the town, and he couldn't have been nicer to me. He told me I made him laugh, and it was an honor to have me work his nightclub. Then he said, "If there's anything I can do for you, just tell me. And I do mean anything."

I said, "Anything?"

He looked me straight in the eye and said, *"Anything."*

Then he told his teenage son to go up to his office, open the top drawer of his desk, and fetch the address book in his drawer. When the kid came back down, the boss flipped through his book and then seemed upset. He told me that he had a special girl he wanted me to "meet," but that she was out of town that night. He said, "I'll have a girl for you tonight who's really nice. But Ella Mae—that's the one I want you to meet—will be here tomorrow night."

I thanked him, and that night a very lovely young lady paid me a visit. On a scale of 1 to 10, I'd give her a 9.5.

The next night was my opening night. I was under a lot of pressure now—this guy had been really nice to me, so I wanted the show to be great. Fortunately, it was, so after the show I got to meet Ella Mae.

She was dressed in farm-girl attire, and she was *hot*. This girl was so hot that if she smiles at you as she turns you down, you think you did all right.

I said to Ella Mae after observing her physical attributes, "You're just oozing sex. I guess when a guy's with you he comes quick."

Then she said, "A lot of them tell me, 'Don't move!'"

I like southern girls. They talk so slow that by the time they say no, I made it already.

Chapter **Five**
I Needed $3,000
to Get Out of Jail

The other night, I had a date with a manicurist. We went to a nightclub. We started to hold hands. And while she was holding my hand, she took my other hand and put it in my drink.

At twenty-eight, I decided to quit show business, get married, lead a so-called normal life. To give you an idea of how well I was doing at the time I quit, I was the only one who knew I quit.

I married a singer named Joyce Indig, who also gave up the business in hopes of having a normal life. We quickly found out that married life was at least as tough as show business.

I later learned that it wasn't show business that was crazy—it was me.

We got no wedding present from my mother, which was no surprise, because she had always hated and put down any girl I liked. I think that was mainly because she wanted to make sure I'd always support her. When I told my mother I was going to marry Joyce, she looked at me as though I had betrayed her, and then made me promise that I would always support her.

After Joyce and I were married, I tried to start a nice little family ritual. I decided that we'd take my mother out to dinner every Sunday. That lasted for two Sundays. All through dinner, my mother looked at Joyce with such hatred that from then on, it was just Mom and me having dinner on Sunday nights.

I tell ya, it's tough to save a buck. Right now I'm supporting two fighters. My wife and her mother.

Being married, trying to start a family, I now needed a steady income, so I went into the home improvement business. I sold aluminum siding and paint on a commission basis at a place called Pioneer Construction in Newark.

I was doing well, but then the weather turned cold and made it difficult to get jobs in the Northeast, so a couple of us decided to go down to New Orleans to work. I took two guys with me as canvassers and covered everybody's expenses.

We did pretty well down there. We worked around New

Orleans for two months, then decided to head back to New York. Driving back, we passed through Birmingham and I said to my guys, "Hey, let's stop here for a few days, give it a shot."

After we got a hotel, I approached a siding company that looked like it was reputable. I asked the owner, a guy I'll call Steve McGill, if I could be a sales rep for him. He said, "Okay, fine."

My two canvassers and I went to work the next day. We worked hard, and everything was going great—we were signing up a lot of customers for McGill. After a while, though, I noticed that none of those jobs were actually getting done, so I said to McGill, "What's happening? I got about fifteen jobs out there, but you're not doing them." This was a problem for me because I didn't get paid my commission until the job was finished.

"I'll get to them when I get to them," McGill said. "I'm pretty busy right now."

Yeah, and I was busy, too—paying for three hotel rooms, my car, my expenses . . .

"All right," I said, "as long as you get to them. I mean, they're three weeks old. You start 'em and don't finish 'em, it really aggravates people."

"I'm doing the best I can," he said. "About how much you got coming from me?"

I said, "About four thousand dollars."

"Look," he said, "if you want to end this thing right now, I'll give you a check for seven hundred and fifty, and we'll call it even."

I was stunned. "In other words," I said, "you want to burn me out."

"That's my offer," he said coolly. "If you want the seven-fifty, okay. If not, I don't know when I'll get to those jobs."

I was really disgusted. I saw no way out.

That was on a Saturday afternoon. That night, I went to a nightclub. I wanted to have a few drinks, see a floor show or a comedian, maybe forget for a few hours that I was getting screwed by McGill. So I fall into this place, say hello to the owner, and after a few drinks, I feel a little better. In fact, I feel like getting up and telling some jokes to break the mood.

I tell the owner I'm a comedian, and I'd like to do about five minutes, no charge. He says, "Go ahead," so I got up on the stage. I was half loaded, but I'd been performing for at least ten years at that point, so I knew what I was doing. I did about seven or eight minutes, and got a big hand as I walked off.

I'm sitting back at the bar now when the boss comes up and says, "There's a fellow sitting over there with his wife who'd like to say hello and buy you a drink."

I said, "I don't know, man. I'm not feeling real friendly tonight."

"Well, he's a helluva nice guy," the boss said. "What can you lose?"

I said, "Okay, what the hell."

So I have a drink with this guy and his wife. The guy said to me, "You're a funny man."

I said, "Thank you very much."

He said, "But you look like you're worried. What's the matter?"

I said, "No one can help me with what I'm going through."

"Come on," he said. "What is it?"

So I told him about McGill.

This guy listened to the whole story without saying a word, then said, "Maybe I can help you. Don't say anything to McGill, but call me Monday afternoon." He gave me his card. His name was Al Fontaine.

It seemed crazy, but I had nothing to lose, so Monday afternoon I called him and he said, "Everything is taken care of. McGill will have a check waiting for you."

I thought to myself, *I don't believe it!* But I went over to McGill's, and sure enough, he gave me a check for the full amount—$4,000.

It turns out that my buddy from the nightclub was a big shot with all the local banks because he ran the largest home improvement business in Birmingham.

First thing Monday morning, he called the bank where McGill did business and said, "Look. We don't want guys from up north coming down here and taking jobs from our salesmen. McGill is working with a guy from New York, a real go-getter, and I want to get rid of him. I think it's best if we tell McGill to pay this guy off immediately and get him out of town."

McGill didn't know what hit him. How did Jack Roy, some stranger from out of town, get to the head man of *his* bank? It was a mystery that would never be solved.

Before I left town, I bought the most expensive sweater I could find, and took it over to the office of my new best friend, Al Fontaine. I gave him the sweater, and a big kiss on the forehead.

That was the first time I kissed a guy—we'll forget about that guy who'd give me nickels to sit on his lap when I was five—but I felt that this guy had worked a miracle.

I tell ya, southern people, they always think you are hard-of-hearing. Every time you leave they say to you, "You come back, you hear?" And southern people, they think you are horny, too. You get directions, they say, "Just up the road apiece."

At this point in my life, I hadn't seen my father in twenty years. I knew that he and my uncle Bunk had moved to Los Angeles, so I decided that my wife and I would take a train ride out to the West Coast to see him.

I was very curious to see how my father was doing at that stage in his life, how he'd adjusted to not being a performer anymore. I knew he had gotten out of vaudeville and made some good money playing around on Wall Street. That had been a long time ago, but I thought, *Who knows? He's liable to be in some Beverly Hills mansion.*

The trip took five days. It was an enjoyable ride with

the rumble of the train, playing cards, looking out at all those honey-colored wheat fields.

When we got to Los Angeles, I looked up my father's number and gave him a call. He told me to meet him at his job later that day.

When I got down there, I was shocked to see that he was working as a salesman at a Karl's shoe store. I didn't know what to say, so I didn't say a thing. Here was this guy who had been a top stockbroker, knew everybody, and now he was restocking racks of shoes.

Not that my father and I had ever been close, but that really got me down. I thought seeing my uncle Bunk would cheer me up, so I got his number from my father and called him.

My father warned me, "Bunk is not himself."

"I don't care," I said. "I want to see him. I'm just gonna say hello and leave."

So I go to see Bunk, who I hadn't seen in twenty years. I ring his bell, and he hollers, "Come in!"

I walked into his bedroom. He was sitting up in bed watching television.

I said, "Hey, Uncle Bunk, how ya doing?"

He just kept looking at the TV as he yelled at me, *"Sullivan! Sullivan!"*

I figured I'd wait till Sullivan was over. Then I said, "Hey, Uncle Bunk—"

He cut me off again. He said, "The news! The news!"

I didn't have time to wait around until that TV station went off the air, so I decided to go back to my hotel.

As I walked out, I yelled to Uncle Bunk, "Hotel! Hotel!"

In L.A., as usual, my father had a lady friend. He told me in his sixties there were plenty of women available whose husbands had died. He was going with his L.A. girlfriend for about a year when he decided to take a trip to Florida.

My father's girlfriend said, "Phil, you must stop by and see my girlfriend in Texas. I've told her so much about you."

He looked at her picture and said, "Okay, I'll say hello when I'm passing through."

My father pulled up in her driveway. It was a lovely house and a nice town. He stayed with her for four years—then he continued his trip to Florida.

With my ol' man, I got no respect.
He told me to start at the bottom.
He was teaching me how to swim.

My home life wasn't anything to brag about either. In 1949, after Joyce and I had been married for ten years, she gave birth to our son, Brian. Six months later,

LEFT: *Here's my first wife, Joyce, on our anniversary, still giving me the cold shoulder.*

Joyce and I got divorced. I loved my kid, but I couldn't stand my marriage.

While my home life was falling apart, I started getting closer to my father. I would occasionally fly down to Florida and spend a few days with him. Of course, my father had a lady friend down there, but now that he was in his seventies his priorities had changed. Now he looked for only one thing in a woman—she had to have a car.

One day I was driving around with my father down in Miami. We came to a red light. All of a sudden my father says to me, "*This* is the place to live."

I looked around and didn't see anything special, so I said, "Why?"

He said, "Well, there's two supermarkets right over there, and back a block, there's two more."

Elderly people who live in Florida have the same problem every day—how to kill time. There was a barber down there who had a sense of humor. He put a sign in his window. ONE BARBER. PLENTY OF WAITING.

Getting older is tough. I remember the last time I felt an erection. It was at the movies. The only trouble is, it belonged to the guy sitting next to me.

I was glad that I'd gotten a little closer to my father. I wouldn't say it made up for all the years he'd basically abandoned me as a kid, but it was something. I guess that underneath it all he wasn't a bad guy. Even though he had walked out on me, still and all I understood him, and there was a part of me that liked him. Knowing my mother, what were his alternatives?

My old man was really old by now, but he was still pretty sharp. One time I said to him, "You've traveled all over the country, must have slept with hundreds of women. You've done everything, been through it all. What's life all about? What's the answer?"

He twirled his cigar and said, "It's all bullshit."

You can't fully appreciate that line until you're old.

Women my age just don't turn me on. That's
another problem with getting older. I took out
an older woman the other night, and I mean old.
I told her, "Act your age." She died.

Even though I wasn't in show business anymore, I was still making people laugh and, in a roundabout way, getting paid to do it. I learned that in the aluminum-siding business it's a big plus if you can make people laugh. In any group, professional or social, people usually gravitate toward the guy with a sense of humor. I learned pretty quickly that I had a much better chance selling a job if I could make the people like me.

Eventually I went into business for myself.

I was doing pretty good, but after about a year, I found out that my accountant was doing something illegal with my books. How did I discover this important bit of information? The FBI knocked on my door at about six one morning, unannounced. They arrested me and took me and ten other guys to the courthouse in downtown New York.

Bail was $3,000—which I didn't have—so I called my mother. I knew she had the money—because I'd given it to her. Over the years, I'd steadily been kicking her some dough for her savings.

When I told her my story, and that I needed $3,000 to get out of jail, she said nothing. I couldn't believe it.

Finally she spoke. "What am I gonna do?" she said, sounding real pained. "I guess I'll have to give it to you."

"Ma, forget about it," I said. "I'll be all right. You don't have to give me anything."

Suddenly she felt a lot better.

Next thing I knew I was in the backseat of a shylock's new Cadillac, where I made a deal. Now I had the money to cover my bail.

When the case went to court, the judge said I had to pay a fine, and he put me on probation for a year. I had to report once a month to the parole officer. After we'd done that for a few months, I said to the guy, "Do me a favor, will ya? You see I'm a legitimate guy. Let me call you on the phone so I don't have to come down here every month." He agreed, so I did him a favor. I told him how to get girls. I said, "Hang out at women's prisons and wait for parolees."

I sold aluminum siding for twelve years. I made a decent living, but I wasn't living. I was out of show business, but show business wasn't out of me, so I did the only thing that made sense—I created a character based on my feeling that nothing goes right.

Nothing goes right. I joined Gamblers Anonymous.
They gave me two-to-one I don't make it.

Chapter **Six**
Why Didn't You Tell
Me You Were Funny?

*We were poor. We were so
poor, in my neighborhood the
rainbow was in black-and-white.*

S how business was my escape from life. I had to
have it. It was like a fix. I needed it to survive.
I had gotten out of show business when I was
twenty-eight, but if you think funny, you can't just
turn it off. I was no longer performing, but when I'd see
something funny, I'd write it down. I kept writing things
down and throwing them in a duffel bag. After twelve years
in the siding business, that bag was bursting with jokes.

I was forty years old now, and all my friends thought I
was insane to get back into clubs, but I was writing com-
pulsively, trying anything to get a job.

My first move was to go to a nightclub I had worked my
first time around. It was the Miami Club in Staten Island.
The club still had the same owner, Frank Santore, who had
known me when I had worked there fifteen years before.

I had a drink with Frank, who knew that I had quit show business, and told him I'd like to go on that night and tell a few jokes. He said sure.

When my spot was over, Frank said, "Jack, are you taking dates?" Man, that made me feel good. For a few hours, I was able to forget about my troubles.

I was back—hell, I was flying, man—happy to be doing what I knew I wanted to do. Now all I had to do was get some more paying gigs. I went to see an agent named King Broder, who booked most of the small nightclubs on Long Island. Performers referred to these joints as "toilets," but they were places to work, so they looked good to me.

But Broder wouldn't book me. He said, "I don't know you. I'd look ridiculous sending a forty-year-old nobody to a club."

So I started working on him, making him like me—just like when I was a salesman. I said, "Okay, I see your point. I tell you what. On Saturday night, I'll just ride around with you while you check out all your comics in the clubs. We'll keep each other company, see what happens, all right?" He said okay.

Come Saturday night, we hit the first club, and Broder's comic is dying. He's up there maybe ten minutes, and I'm sitting with Broder at a table, and I say, "Jeez, let me go on after him. What can you lose?"

Broder says no. "If you go on after him, I can't win. If you do bad, it makes it worse for me. And if you do good, then the owner will be angry at me for not booking you. You're not going on."

I hated to admit it, but I could see his point. So we go to the next club, still just hanging out together, riding around, doing this and that, whatever business he has to take care of. After a few hours, we eventually walked into a club, and the owner comes running over, furious, yelling at Broder, "Where the hell are they?" Turns out the dance team that Broder'd booked there hadn't shown up.

This sounds like another showbiz story, but, again, it's true. I say, "Let me go up. I'll do ten minutes for you, okay?"

Broder had no alternative, so he says, "Go ahead."

So finally I was able to get on a stage for Broder, and I killed that crowd. Even better, I got hired to work that club starting the next weekend.

And that's how I got back into show business.

With my wife, I got no sex life. Her favorite position is back-to-back. Oh, one night we tried something wild. She tied me to the bed. Then she put her clothes on and went out.

This time around, I was desperate to write jokes and go out and tell them, and I went to extreme circumstances to do it.

I was always trying to get work in the Catskill Mountains,

a resort area about ninety minutes north of New York City, even though it was really not my kind of gig. I was brought up in nightclubs, and the Catskills were mostly hotels and small resorts, so the comics who played there were more into family entertainment, but I could write material that would play there, and I was trying to get any work I could get. An agent named Hy Einhorn booked a lot of acts in the Catskills, and he would book me into the small hotels or bungalow colonies if he had no one else.

I called Hy one day and asked if there was any work for that night. He told me there was one hotel in the Catskills that might have a show. They wanted to wait to see how many guests checked in, though, so he told me to check back.

I called him an hour later. He said, "They don't know yet."

Another hour goes by. I'm sitting there ready to drop everything and drive an hour and a half up to the Catskills just so I can get up onstage and tell jokes.

I called him again. No word.

Finally, at about five o'clock, he said, "Nothin' doin'. They didn't have enough check-ins. There's no show."

I said to him, "Hy, do me a favor, will ya? The hotel would pay me twenty-five dollars for a show, right? I don't want the twenty-five dollars. You'd make ten dollars commission out of that. Skip your commission, which you wouldn't be getting anyway. So the hotel just pays for the dance team. They get fifteen dollars, right? Call the owner, tell him you'll give him a complete show for just fifteen dollars. I get nothin', you get nothin'. You're doing him a favor, and you're doing me a favor."

He thought about it and then said, "All right."

So my job that night was to pick up the dance team— a guy and his girlfriend who lived in a hotel downtown— and drive all of us up to the hotel.

When we finally arrived at this third-rate Catskills hotel, I went into the canteen to buy a soda. I introduced myself to the woman at the cash register. "Hi, I'm the comedian for tonight's show," I said.

This woman looked at me like she hoped it wasn't true, and said, "*You're* the comedian? Hy must have really been stuck."

I found out later that she was the wife of the guy who owned the place. Within a half hour, she'd told everybody at the hotel not to expect too much from that night's show because the comedian was a stiff.

I didn't know this at the time, of course, so come showtime, I walk onto that stage and the audience is sitting on their hands, looking at me like I was the minister at a wake. It took me a few minutes to warm them up, but I had my act together, and I knew what I was doing. The show went great.

After the show, the boss's wife came running up to me and said, "Why didn't you tell me you were funny?"

*I tell ya, I know I'm ugly. My proctologist
stuck his finger in my mouth. Ugly! Four gay
guys saw me and went straight. Halloween,
I open the front door . . . kids give me candy.*

Being divorced didn't make me much happier, and seeing my little baby son Brian hanging on his milk bottle really got to me. Joyce was pushing for us to get back together, so finally I said yes. We got married again. A year later, my daughter Melanie was born. But the marriage was still not working, so I moved out of the house a short time later.

Despite my troubles at home, I was focused on my act. In fact, I eventually worked up a complete set of material *just* for the Catskills, forty minutes of dynamite stuff. It was so good, I made a deal with a comedian named Stan Irwin, who could get booked in the Catskills where I couldn't get a job. I rented him my act for the summer for two hundred dollars.

That deal worked out pretty well for both of us. Years later, Stan was the producer of Johnny Carson's *Tonight Show* for a while, and when Stan told Carson about renting my act, Johnny got a big kick out of it.

I know I'm ugly. I went to a freak show.
They let me in for nothing.

Another time I hustled my way into a job at the Club Safari on Long Island. I went out there with my friend Joe E. Ross, who was a pretty big comic at that time—he was even starring in a TV show. Some of you older readers might remember him from the *Sergeant Bilko* series and *Car 54, Where Are You?*

The way it worked at the Club Safari, the emcee would introduce Joe E., who was sitting at a table, having a drink. Joe E. would stand up and take a bow, then do a few minutes. I told him, "When you finish your bit, introduce me before you sign off."

So Joe E. told a few jokes, and then I got up, did my act, and again my plan worked—I busted up the joint, and got booked there for the following weekend.

After I came offstage, I went back to my table and sat with Joe E. and a lady friend I'd brought along. The band started playing, and she wanted to dance. I told her, "Sorry, honey. I'm a bad dancer. Dancing is not my thing."

Joe E. said, "Come on, baby, I'll dance with ya." So off they go.

After a while she came back to our table and Joe E. went to the men's room. When she sat down, she told me that while they were dancing, Joe E. had said, "Baby, I'm

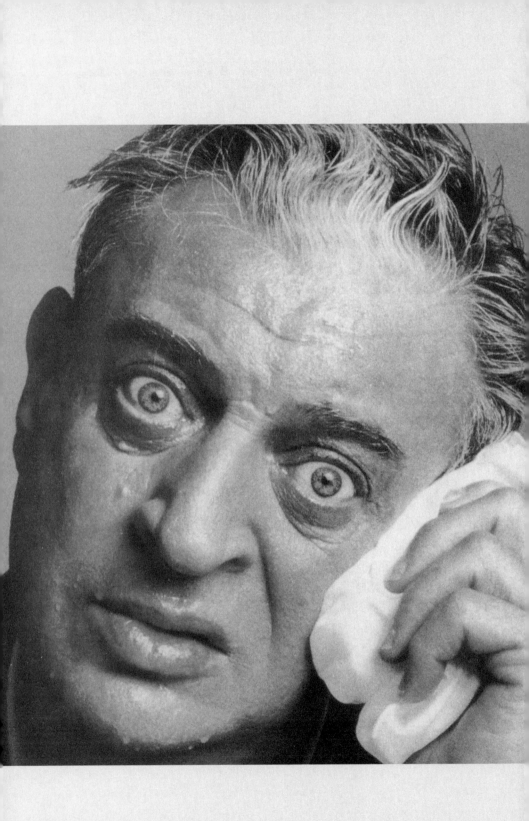

doing better than he is. Pack your bag and come home with me."

Note that "bag" was singular. In most of Joe E.'s relationships, the girl only needed one bag. It wasn't going to be a long "trip."

As you might have figured out by now, Joe E. was a rogue. He also had a habit of marrying hookers. When he had an argument with one of them, he'd say, "Go fuck somebody!"

Once Joe E. was working a nightclub with an act called the Dancing Paceys. Both Joe E. and I were friendly with Jimmy Pacey, a wild, wild guy.

One night a sea captain and his wife came to the club. They drank and had a great time, so the sea captain invited Joe E. and Jimmy to his house for a nightcap. The two thespians said, "Okay!"

Before they left, Pacey fucked the captain's wife, and Joe E. stole the captain's overcoat. The moral of this story: Never invite a comic and half a dance team to your house.

Joe E. had a thing about coats. One day he and I were walking in New York. We came upon a huge men's clothing store called Howard's. Joe E. said, "C'mon. I need an overcoat."

When we were inside, Joe E. waved off the salesman. "I'm just lookin'," he said.

I sat down and watched Joe E. try on overcoats. When

LEFT: I tell ya, I know I'm ugly. I stuck my head out the window—I got arrested for mooning.

he put on one that he really liked, he got that look in his eye that I knew meant trouble. He said, "Let's go. I'm walkin'."

I didn't want to get caught stealing a coat, so I said, "You're walkin' alone." I went out about twenty feet behind him . . .

That maniac got away with it—he stole that overcoat.

We walked over to Joe E.'s hotel to tell his hooker wife the good news. While Joe E. was telling her the story, he farted. At that, his wife yelled at him, "Joe! You ate and you didn't bring me anything?"

I figured out I'm bisexual. I have sex twice a year.

At that time, there was another famous comedian called Joe E. This guy, Joe E. Lewis, was really big—much bigger than my friend—and he had a special song written

RIGHT: I was playing Upstairs at the Duplex in Greenwich Village. I had been drinking and was feeling pretty loose. I used to end my act with "That's my story." That night, I said to the audience, "Forget the applause, all right, when I walk off, just give me one of these"—and I formed a circle with thumb and forefinger. It became my signature exit line. If you like the book, give me "one of these."

for him called "The Groom Couldn't Get In." It was one of his best numbers, and he used to close his act with it. Joe E. Ross liked that song so much that he stole it, and started using it in *his* act.

One night the two Joe E.s ran into each other in a restaurant. Lewis said to Ross, "What you did was very low. You stole my number."

Ross said to Lewis, "I'll make a deal with you. Let me keep doing the number, and you can keep fucking my wife."

I never got girls when I was a kid. One girl told me, "Come on over, there's nobody home."
I went over. There was nobody home.

One night Joe E. and his hooker lady friend decided to try to live like a normal couple. The idea was that she'd stop turning tricks and he'd stop chasing everything in a skirt.

The next afternoon, Joe E. was sitting in the lobby of the club he was working at, across from the boss's office. Suddenly the boss's door opens and out walks a midget. About two minutes later, Joe E.'s lady friend walks out of the office. It was obvious what had just taken place.

Joe E. grabbed his girl's arm and said, "I thought you were going to straighten out—no more turning tricks!"

She said, "Joey, he's a midget." She smiled and snapped her fingers. "He came like that."

I heard that the circus had a problem with a midget. The midget walked over to the fat lady and said, "I wanna screw you." She said, "If you do, and I find out about it, you're in big trouble."

One night Joe E. and I were drinking in the bar of the Havana Madrid nightclub in New York. Joe E. was loaded, and I wasn't far behind. A girl walked over to him and said, "Hey, Joey, how are you?"

Joe E. put his arm around her and said, "Yeah, baby, come here. Tonight it's you and me."

She said, "Joey, I'm your sister!"

Joe E. was slightly embarrassed . . . but he also looked disappointed.

I tell ya, yesterday was a tough day. I found some guy's wallet. Inside was a picture of my two kids.

Chapter **Seven**
Some Show Business on the Side

I hate blind dates. The last blind date I had, the girl was ugly. Only once in her life was she whistled at. It was right before the train hit her.

was doing better as a comedian the second time around. I was older and wiser, yeah, but I was funnier, too. I was really working hard on my jokes, and polishing all that material I'd stored in that duffel bag for twelve years. My timing was better, my jokes were better, and my name was better. Yeah, I was no longer Jack Roy.

One of the biggest changes I'd made in my act was my name.

Early in my comeback, I visited a club I'd worked at years before, hoping I could get booked there because I'd been one of their favorite comics. I hadn't worked there in quite a while, though, and the club had new owners, but they knew of me from the people who'd been coming in there for years. I talked to them, and they finally booked me.

At that time, if you were working in a club, they'd put your name in Friday's edition of a newspaper called *The Mirror*. There were hundreds of nightclubs at the time, and the Friday *Mirror* had the names of all the acts and where they were working.

I said to the owner, "Do me a favor, will ya? I haven't worked in a long time, and I don't know how I'll do, so put a different name in *The Mirror*. Any name at all. Just don't put in Jack Roy, okay?"

He said, "Okay."

So he makes up a name and runs it in the paper.

Despite my attempt to perform "anonymously," word got around the neighborhood that I was appearing there, and plenty of people who'd dug me years ago showed up, which led to some confusion. When it was time for me to go on, the emcee said: "Here's Rodney Dangerfield."

I walked out on that stage and it felt weird. I saw all the same faces, only now they were twelve years older. And they looked at me, then looked at one another, and said, "Rodney Dangerfield?"

I said, "Hey—if you're gonna change your name, change it!"

My show went fine, despite my nervousness, and afterward I asked the owner, "Where'd you get that name?"

"I don't know," he said. "I made it up, just like that."

All my friends said it was a funny name, so I decided to keep it. My wife told me, "With a name like Rodney

Dangerfield, if you don't hit, you're an idiot." She said I should write a bit about my new name.

One day I had nothing to do, so I gave it a try. I wrote it in one afternoon.

This is jumping forward a few years, but a while later I made an album called *The Loser*. It became popular in England because of the bit about how I got my name. I called it "What's in a Name." It went like this:

When I went into show business, I saw an ad in the paper. It said: "Improve Your Personality." So, I went to see the man.

He told me my personality was okay, but my name was my problem.

I said to him, "My name? How could a name be a problem? Even William Shakespeare said, 'What's in a name?'"

He said, "Who?"

I said, "William Shakespeare."

He said, "Look, do you want to listen to me or do you want to listen to your friends?"

I said to him, "I don't understand. Is it good to change your name?"

He said, "Of course. I always keep changing my name. In fact, now I can give you a very good deal. I have a new name coming in next week, and I need the space. I can give you a new name for five hundred dollars."

I said, "Five hundred dollars? That's a lot of money."

He said, "It's a great name. It's a name once people hear it, they'll start saying it."

I said, "What's the name?"

He said, "Rodney Dangerfield."

I said, "Rodney Dangerfield?"

He said, "See, you just heard it, and you're starting to say it! Listen to me, take the name."

I said, "Wait a minute. Suppose I use the name and I don't like it. Can I bring it back?"

He said, "Of course. All I ask is one thing. While you're using the name, don't give it a bad name!"

So I decided to call myself Rodney Dangerfield. As soon as I got home, I thought to myself I made a mistake. I called the guy up. I said, "Look, I want my money back. This is Rodney Dangerfield."

He said, "Who?"

I said, "Dangerfield! Don't you remember?"

He said, "Oh, yeah, Shakespeare's friend."

I said, "Look, I don't want the name."

He said, "Don't be foolish. You have to get used to it. Sit in hotel lobbies, have yourself paged. Try it for two weeks, I guarantee you'll like it."

So I tried the name for two weeks. I still didn't like it. I went to bring it back. I couldn't find the guy.

He had changed his name.

RIGHT: My stock publicity photo when I reentered show business.

Around this time I bought a new car and I picked a manager for a strange reason. I was still doing a lot of one-nighters in the Catskills, and after my show I liked to get drunk. It was about a ninety-minute drive back down to New York, and I was usually in no condition to drive myself home, so I started looking for a manager who was a good driver.

In 1963, after a couple of tough years into my comeback, I got a big break, a chance to audition for *The Ed Sullivan Show,* the biggest variety show on television back then. I went on in the afternoon, after the dress rehearsal. I followed Dame Judith Anderson doing a death scene from *Macbeth.*

I can still remember some of the jokes I did that night:

> *I live in a tough neighborhood. When I plan my budget, I allow for holdup money.*

> *I tell ya, in my building, nothing but robberies. Every time I close a window, I hit somebody's hands.*

The Ed Sullivan Show audition was a tough test, but I was rehearsed and ready, and everyone said that I had done well. Now I had to go home and sit by the phone, waiting to see if Sullivan would book me on his show.

Three weeks went by and I heard nothing. Then I got the call. He booked me on the show for March 5, 1966, for $1,000. I was broke—and the happiest guy in the world.

When that big night came, I remember sitting in my dressing room, waiting for the show to start. I looked out the window. It was raining, but the streets of midtown Manhattan were crowded and I thought to myself, *Look at all those people who are gonna miss seeing me tonight on* The Ed Sullivan Show.

My bit went great, and they booked me for a second show at $1,500. That second performance also went well. I was finally getting somewhere.

Time and tide and hookers wait for no man.

One night when I was doing *The Ed Sullivan Show* I lost my place. What happened was I didn't sleep the night before. I was okay at the dress rehearsal in the afternoon. Then at night, just before the show started, I was exhausted. I couldn't keep my eyes open. I walked out and did three or four jokes, and all of a sudden I thought, *What's next?*

I just kept saying, "I don't know. What can I tell ya? What can I tell ya?"

I looked at Sullivan. His face looked like he knew something was wrong. I figured he knew I lost my place. I felt terrible.

I was going through torture trying to think of the next

joke. Now, in my head I am groping to find any joke, any joke at all, and finally I thought of a joke. I told it. Then I realized I jumped about three jokes ahead. I just lost those and went with the rest of it.

I took my bow and went over to Sullivan. He said, "Ladies and gentlemen, Rodney Dangerfield." Then he turned to me and said, "If there's any more like you at home, bring them over here." I was surprised. Then I went backstage and everyone told me how good I did. No one even realized that I lost my place. But me, I'm too strict with myself. I'm still not over it.

---

*I tell you, I can't take it no more. My dog
found out we look alike. He killed himself.*

My career was heating up, but I was still selling siding, a strange combination that led to some funny experiences. At that time I was talking to a big agent, Dee Anthony, trying to get him to represent me. Dee was going to visit the great singer Tony Bennett at the Copacabana. He asked me to join him.

After the show, we all hung out backstage, had a few drinks, and had a great time.

I told Tony I had an aluminum-siding business in Englewood, New Jersey, and Tony said, "Hey, that's only about ten blocks from my house. Give me a call sometime. We'll get together, bullshit a little."

I called him the next day, and he said, "Come on over."

So I go to his house, and we're talking about this and that for about an hour when I remembered that I had a siding job I had to check on. I said, "You wanna take a ride with me?" He said okay.

When we got to the job, I said to one of the siding mechanics, "You know who I got in my car? Tony Bennett!"

He goes over to the car and starts talking to Tony.

The other workers see this, and they want to talk to Tony Bennett, too. Next thing I knew, the woman who owned the house came out with a camera. Then the neighbors wanted autographs. It became a circus.

I got in the car and said, "A lot of fun, hey, Tony? We'll do it again sometime."

Since then he don't return my calls.

Another time, I sold a siding job to a couple on a Saturday. That Sunday night, I did the *Sullivan* show.

The following Monday morning, the woman asked one of the guys on the installation crew, "Is Mr. Roy in show business? I think I saw him on *The Ed Sullivan Show* last night. But they called him Rodney Dangerfield."

The guy said, "Yeah, that was him. He does some show business on the side."

Not long after those two appearances on *Sullivan*, I was working at the Copacabana Club in New York when

Ed Sullivan came in with a small group of friends. The Copa show went well, and as I walked off after my act, Sullivan jumped in front of me with a big smile and shook my hand.

"You must do our show!" he said.

"Mr. Sullivan," I said, "I'm already doing your show."

After that, I got four more shots on *Sullivan*. I still had plenty of problems, but I knew I was going to make it in show business.

I told my dentist my teeth were all getting yellow. He told me to wear a brown necktie.

My appearances on the *Sullivan* show led me to come up with another key part of my new act—my wardrobe, if you'd call it that. For the past forty years, I've always worn a red tie, white shirt, and black suit onstage. That happened mainly because I have no taste in clothes, so I'm not real confident when it comes to matching ties and shirts and shoes and all that other stuff you have to worry about if you want to dress nice.

When I did my first *Sullivan* show, I thought, *What*

RIGHT: *Things are tough. Now I'm taking in laundry.*

should I wear? It was my first time on television, and I wanted to look good, so I picked something safe: red tie, white shirt, and black suit.

Worked just fine. Nobody complained about my clothes. Nobody complimented them, either, but . . .

Two months later, I'm on the show again. What should I wear this time? I don't know. So I said, "I'll wear the same thing. Who cares?"

My third shot? Same thing.

Any other television show? Always the same thing.

Black suit, white shirt, red tie. By now it was like a uniform.

One of my outfits is now in the Smithsonian Museum in Washington—right next to Lindbergh's plane. I hope they're not using my shirt to clean Lindbergh's plane.

With me, nothing goes right.
My psychiatrist said my wife and
I should have sex every night.
Now we'll never see each other!

Thanks to my "What's in a Name" bit, I got booked on the biggest TV show in England, *The Eamon Andrews Show.* That was on a Sunday. Then I had to fly back to the States to do *The Ed Sullivan Show* the following Sunday. I thought, *Boy, I guess I'm in show business—Sunday the*

*best television show in England, and next Sunday I'm
doing the biggest television show in the U.S.*

After a few shots on *Sullivan*, it was easier for me to get
booked in clubs, and I was now earning $4,000 a week on
the road. The television talk shows would now put me on,
too—Merv Griffin, Joey Bishop, Mike Douglas—which
made it much easier for me to get booked in bigger clubs,
where I made better money. Then my agent made a deal
with *The Dean Martin Show* for me to appear on twenty-
eight shows. I signed on to do some short skits—just me
and Dean—and I would write all the material.

Dean only came in once a week to tape his show—no
rehearsal. (The set for our bits was always the same—me,
Dean, a table, and two chairs.) For our first show together,
it took Dean and me just three or four minutes to film our
routine and we were done. "Okay, great, see ya next week,
right?" Wrong. That was the last time I saw Dean. For the
next seven Sundays, I flew from New York to California,
went into an empty studio, sat down at that table by
myself, and did four skits while talking to an empty chair.
Later, the crew filled in shots of the audience laughing,
and they filled in Dean Martin, too.

After the taping, it was back to the airport and back to
New York. Many times I thought, *Is this show business?
Doing jokes to nobody, piped-in laughter, no audience?*

I used to have to fly quite a bit back then because I was
doing shows all over the country. I used to tape-record my

act so that I could polish my jokes. When I'd make changes in my act, I'd play the tape back so I could hear how they went over. I'd be sitting on the plane with the earplug stuck in my ear, listening to my act. Nine times out of ten, I'd fall asleep this way.

While I was dozing, I'd hear one of my jokes, and think, *Hey, wait a minute, that's my joke!* Then I'd hear this guy tell another two or three of my jokes, and I'd think, *What the hell? Who is this guy? He took my whole act!*

I'd be so upset that I'd wake up.

Then I'd realize that I was listening to myself.

My old man was dumb. He picked a guy's pocket on an airplane and made a run for it.

You meet all kinds of people when you're traveling. Once there was a guy sitting next to me on an airplane bothering me with all kinds of questions. Then he said, "It must be rough for you with all the people who bother you."

LEFT: *Dean was such a big star that he didn't even have to show up for his own show. Most of my bits "with" him were done in an empty studio. And he was edited in later.*

I said, "Nah, it's all right." Even though *he* was the one who was bothering me.

And he kept doing it, so I finally decided to smoke a cigarette, which was allowed on planes in those days. I took out a cigarette and started smoking it so that I wouldn't have to answer any more of this guy's questions. Just to make sure, I leaned back against my pillow and closed my eyes. The guy looks over at me and he doesn't know what to say—it was the first time he'd ever seen a man smoking while he's sleeping.

With my wife I get no respect.
I fell asleep with a cigarette
in my hand. She lit it.

On one of my flights, I happened to sit behind Vice President Hubert Humphrey's wife, Muriel. We were sitting there for a while when she pushed her seat back a few inches. She thought maybe she hit my leg, she didn't know, and she quickly turned around to ask if I was okay.

I said, "Hey, baby, don't worry about it. No problem."

The guy next to me said, "You called her 'baby'?"

Oops. I felt guilty. I turned first class into low class.

*My dog, that's another one. All day, he
keeps barking at the front door. He don't
want to go out. He wants me to leave.*

Things were really falling into place for me now, but
there was one door I couldn't pry open. The biggest
show for a comic, the one that could make your career,
wouldn't book me. Johnny Carson's *Tonight Show* wouldn't
use me because my manager—the good driver—had
offended Johnny. Without telling me, he sent Johnny an
offensive letter that accused him of using one of my jokes
on his show. The next night Johnny read the letter on the
air, made fun of it . . . and my manager.

I felt sorry for my manager, so I sent Johnny *another*
offensive letter. So you can understand why he didn't want
me on his show.

About two years later, some people from *The Tonight
Show* came to see me at a club in Greenwich Village. (I
remember that the job only paid $50 a week, so I told the
owner to keep the money, and I drank for free that week.)
The booker from *The Tonight Show* loved my act. He
came backstage and told me he wanted to book me on the
show, but I told him about the nasty letters and said that
Johnny might still be angry at me. This guy said, "Nah.
That was years ago—he's forgotten about that." So I get
booked on *The Tonight Show.*

The night before I was supposed to appear, I got a phone call at my office in New Jersey from the man who booked me. He said, "You're right. Carson is still angry. You're not on the show."

But I wasn't surprised. In fact, I understood.

The next time I crossed paths with Johnny Carson, it was a couple of years later, under unusual circumstances. I had worked the Copacabana for a week and had closed the night before. My aluminum-siding pal Tony Bennett was the new headliner, and I'd come down to catch his show, which was sold out. When I got there, the place was a madhouse—there was a crowd outside, people pushing and shoving, trying to get in.

As I'm walking up to the club, I see my old friend Stan Irwin, who's now a producer for *The Tonight Show,* sitting in a car with Johnny Carson. They'd come down hoping to see Tony Bennett but were about to go home because they couldn't get in. I said, "Come on, follow me."

I took them in the back way, through the kitchen, then led them to the main room, where the maître d' took care of them. During the entire walk, I said only five words to Carson: "I'm sorry, I was wrong." And I meant it.

I went on to do seventy appearances with Johnny. Carson did a great straight for me, and he has a great wit. To give you an idea, one time on his show, I did this joke: "Johnny, the most important thing in life is to have

RIGHT: Johnny Carson was a class act. I guess he put me on the show for contrast.

friends, good friends." I pointed to his announcer-side-kick, Ed McMahon. "Like you and Ed are good friends. How long have you been friends?"

Carson said, "About twenty-four years."

And I said, "No children, huh?"

That got a laugh.

Then Carson came back with, "It's not that we haven't tried."

That got a *huge* laugh.

I remember before we were married,
I told my wife, "Honey, I love you.
Will you marry me?" She said, "If you really
loved me, you wouldn't ask me to do that."

Chapter **Eight**
I Am Not High!

*My doctor told me to watch
my drinking. Now I drink
in front of a mirror.*

So, now I'm back in show business. I should have
been a real happy guy, but I wasn't. I still had my
bouts with depression.

Like most people in that situation, I tried to
self-medicate, which is New Age talk for "I got loaded." I
used to drink. A lot. Too much.

When I was drinking, sometimes my judgment was not
at its best. I left a club one night after getting half loaded
and I was *very* hungry, so I drove to a deli called Smiler's
that was open all night. I grabbed a package of Swiss
cheese and asked the girl behind the counter to cut me a
roll. I unwrapped the cheese and put it in the roll, took a
big bite, and away I went.

I got home, parked my car, got up to my apartment,
and I was almost finished eating the Swiss-cheese sandwich,
but I was thinking, *Wow, this cheese is chewy!*

So I looked at it. I was at the end of the sandwich, which meant I had eaten almost all the cheese . . . and all the pieces of wax paper between the slices. No wonder it didn't taste quite right.

That's what I did when I was drunk. If I had been high on pot, it would have been different—I would have eaten the wax paper, the cheese, and the girl who sold me the cheese.

---

*I went to the store to buy some rat poison
and the clerk said, "Shall I wrap it,
or do you want to eat it here?"*

All my life, I've heard people say that booze is a social lubricant, but I always found that people got along much better when they were smoking pot. Back in 1949, I was working a nightclub called the Famous Door on Fifty-second Street. The show was me and a three-piece band, all black guys.

My first night there, the place was empty, but the boss told me to start the show. I said, "How can I start the show? There's no audience."

He said, "You do the show anyway. If someone is walking by outside and there's no show, they'll just keep walking. But if they see that the lights are on and there's a guy onstage telling jokes, they might come in."

He knew the business better than I did. I got up on that stage and told my jokes to the empty chairs and the empty tables, and after about fifteen minutes, people did walk in. So, I did my act again, then introduced the trio. After they finished, I came back, did another fifteen minutes, and closed the show. An hour later, we'd do it all again.

That first night, when the first show was over, the musicians and I were all sitting in one big dressing room, talking about nothing in particular, when one of them said to the other two, "Wanna get some ice cream?"

They both perked up and said, "Yeah, let's go."

The next night, same thing.

Then it hit me. *They don't care about ice cream. They're getting high.*

The following night, between shows, I said, "Hey, boys, I've got some great shit. Wanna get high?"

They looked shocked—in those days most people thought marijuana was only smoked by black musicians—but they quickly got over their surprise and said okay.

After that, it became a nightly thing—between shows, we'd all go out for "ice cream."

I tell ya one thing about me.
I say "No" to drugs. When people ask
me for some of my drugs, I say "No."

first smoked pot back in 1942. I was twenty-one, just on
the edge of getting into show business, working a
Saturday night here and there, just plugging away. One
night in Kellogg's Cafeteria on Forty-ninth Street in New
York, a popular hangout for acts when they finished working, I was sitting with a couple of show folk—a comic
named Bobby Byron and my *Car 54* friend Joe E. Ross.

We were sitting there talking, doing what we were
doing, and Bobby and Joe E. decided that they wanted to
get high. They invited me to join them.

"I don't get high," I told them, "but I'll keep you company."

So we walked back to the Belvedere Hotel, which was
where Bobby lived. I even remember the room, 1411. We
went upstairs, and Bobby took out a joint and lit it. He
and Joe E. took a couple of hits off that joint, then offered
it to me.

"No, no," I said. "I'm not into that."

It took them about ten seconds to convince me to try
it. After a while, Bobby asked me how I felt. I said I felt
fine, no different, and Bobby started laughing at me.
"You're high," he said, "and you don't even know it."

"I'm not high," I said.

He said, "Okay, why don't you stand up?"

I said, "I tell ya, I'm not high, man."

"Okay, then," he said. "Let me see you stand up and
walk."

LEFT: *Nobody partied harder than John Belushi—not even
me. Here's a rare shot of both of us standing up.*

I said I wasn't going to get up, and he said, "You can't, because you're too high."

I said, "I am NOT high!" and to prove it, I stood up and walked around my chair a few times.

That made Bobby laugh even harder. With him, you couldn't win.

"That proves you're high," he said. "If you weren't high, you never would've gotten up."

After a few minutes I realized I must be high. I felt relaxed, peaceful, everything was okay. That night I found a new friend for the rest of my life.

I tell ya, I'm a bad drinker. I got loaded one night.
The cops picked me up. The next morning I was in
front of the judge. He said, "You're here for drinking."
I said, "Okay, Your Honor, let's get started."

Back in those days, we used to call really good marijuana "boss pot." The boss rarely came around, though. He was too busy. He had a lot of territory to cover. But when you did get boss pot, it was like, "Wow!"

Going back about fifty years, I was with a friend when a guy told us, "I have a pound of boss pot."

Everything stopped. You didn't hear that very often—the "boss" could be elusive—so I bought some from him.

As soon as I got home, I smoked it, and it really was the boss. So much so that I had a problem—I was working in a nightclub called the Golden Slipper out in Glen Cove, Long Island, and I had to do my act that night. The problem was that I can't do my act when I'm high on pot because my timing is all off.

I smoked that boss pot at about six or seven o'clock, then got in my car, drove out to the club, and had dinner. I knew that I usually came down from a high when I ate a big meal, so I wolfed down everything on the menu, but the food didn't dent my high at all. The show was now only a half hour away.

I got dressed and was ready to go out onstage, but I was still stoned—whacked, in fact. I was thinking, *What am I gonna do?* I then remembered that I could work pretty well on a few drinks—so I decided that the best thing I could do was to counteract the high by getting drunk real fast.

I went to the bartender and said, "Line up four shots of Scotch and a beer, will ya?"

He said, "What's with you?"

"I know what I'm doing," I said. And I knocked down the booze and the beer—*bang, bang, bang*—to put my head in a different place . . . which it did. Just not the place I was looking to go to.

The show started, I did my act, and it was very tough for me. The pot was stronger than the booze. My timing was terrible.

After the show I called the guy who had sold me the

boss pot. I wasn't mad at him . . . I wanted to buy some more.

---

*I solved my drinking problem. I joined
Alcoholics Anonymous. I still drink, but I use
a different name. Oh, when I'm drinking,
I don't know what I'm doing. Sometimes the
next day, I wake up in a strange bed, with
a woman I can hardly remember and a kid
with an accent playing with my feet.*

All the stories you hear about people getting wild on marijuana are ridiculous. It's been proven that pot does not make you violent. In fact, it makes you passive. When you're high, the last thing you want to do is fight.

Booze is the real culprit in our society. Booze is traffic accidents, booze is wife beating. People see a picture of a cocktail glass and they think, *How dainty, how sophisticated.* They oughta think about what booze leads to—you lose your wife, your home, your life. In my life I've seen many doctors and psychiatrists, and all of them have told me that I'm better off with pot than with

RIGHT: Here I am receiving one of the greatest pieces of mail I ever got: Dr. Bearman's recommendation to smoke more pot.

booze. In fact, I now have written authorization from a California doctor that allows me to smoke pot for medicinal purposes. It's a license to get high. What better reason to move to L.A.?

Wish I'd had that "prescription" thirty years ago; life would have been easier. I was sitting in an airport one time, waiting for my plane. There was no one around, so I lit up a joint. I was taking a few tokes from it, but no one noticed, because it was a busy place. Everything was cool—or so I thought.

Suddenly a cop came running toward me. I had the joint in my mouth, so I took it out of my mouth with my left hand and let it hang down by my side. When the cop said, "Hey, Rodney!" I figured I was screwed.

But when he was standing right in front of me, he said, "Rodney, can I have your autograph?"

I said, "Sure."

He had paper and a pencil with him. "Just make it out to Fred," he said.

I said, "No problem."

He gave me the pencil and paper, and with one hand— my right hand—I wrote: *Fred, good luck. Rodney Dangerfield.*

As he walked away, I put the joint back in my mouth and took another hit.

Don't try that unless you're in show business—and out of your mind.

I tell ya, my wife and I, we don't think alike.
She donates money to the homeless,
and I donate money to the topless.

When you're high, you become an avid reader. I remember one night I smoked some pot, then started reading the newspaper. An hour later, I said to myself, *What am I doing?* I was reading about fishing conditions in Anchorage. And I don't even fish. And the paper was a month old.

Pot does it for me, but some people want to go further. They try coke.

I did coke for a while. What a mistake that was. Coke is easy to start, and hard to stop. If a group of guys are hanging around and one guy is doing coke, he'll say, "Take a hit. You'll feel like a new man." He's right; the problem is that once you feel like a new man, that new man wants a hit so that *he* can feel like a new man. And that goes on and on until the coke runs out, and you're broke.

When you're on coke, things can be going bad and you think you're doing great. I remember one night I was playing dice in Vegas, high on coke. I had lost $3,000—but on coke, I thought, *Man, I'm doing great! I'm still here!*

Coke makes you do stupid things. One night I was home alone and decided to snort some coke, then watch TV. So I took a shower, then sat down on the edge of my bed and poured out two lines of coke. I snorted them, then sat back on the bed, put my feet up, and turned the TV on.

A couple of minutes later, I spotted some coke down by my feet, at the end of the bed.

I thought, *How did I miss that?*

So I grabbed my straw, sniffed it up, and sat back to watch some TV.

I was sitting there for a while, my feet on the bed, and I saw some more coke that I'd missed.

Now I was thinking, *What the hell's going on?*

Then I realized what was going on—I had powdered my feet after my shower. I had been snorting foot powder.

I wish I could say that was the stupidest thing I ever did on coke.

With my wife I could never have a good time.
The other night I was drinking. She said,
"I want you to stop. You're drunk enough for me."
I told her, "I'm never drunk enough for you."

You do things on coke you wouldn't normally do, and you say things you wouldn't ordinarily say, like, "Honey, I love you. It's you and me against the world forever."

The next morning, you're beat, your heart's racing, you can't breathe, and you feel terrible. You want out, and you hate everything—especially *her.*

One night I was with a chick, and we snorted coke all night. The next day we took a walk on the beach. As we were walking along, she said, "Rodney, did you mean all those things you said to me last night?"

I looked at her. I said, "Who are you?"

I'm not a sexy guy. I went to a hooker.
I dropped my pants. She dropped her price.

Chapter **Nine**
Can I Have Your Autograph . . . and More Butter?

Same thing with my wife, no respect.
I took her to a drive-in movie.
I spent the whole night trying to
find out what car she was in.

People often ask me how I came up with my "no respect" line. When I got back into show business in 1961, I felt—for obvious reasons—that nothing in my life went right, and I realized that millions of people felt the same way. So when I first came back my catch phrase was "nothing goes right." Early on, that was my setup for a lot of jokes.

It worked pretty well, but a few years later, the book *The Godfather* came out and it was a bestseller and then the movie came out, which was even bigger. Because of *The Godfather,* suddenly all anybody would talk about was "respect." *You've gotta show me some respect . . . If she's with me, you show her respect . . . It's all about respect . . ."*

I realized that fit pretty well with the image I was now working with onstage, so I decided to come up with a joke that had the word "respect" in it. The first one I wrote was: "I get no respect. When I was a kid, I played hide-and-seek. They wouldn't even look for me."

I was working a place in Greenwich Village, Upstairs at the Duplex, when I first did that joke, and the crowd loved it. After the show, people came up to me and said things like, "Hey, Rodney, me, too—no respect."

So I kept writing jokes off that line—"I don't get no respect"—and it caught on more and more. Now I have probably written over five hundred "no respect" jokes. Here are a few of my early ones:

With my old man, I never got respect. I asked him if I could go ice-skating on the lake. He told me to wait till it gets warmer.

And the first time he put me on the roller coaster, he told me to stand up straight.

With girls, I don't get no respect. I went out with a belly dancer. She told me I turned her stomach.

When I was a kid, my yo-yo, it never came back.

LEFT: *These days, every underdog—from sports to politics—says they're the Rodney Dangerfield of their whatever. Two years ago, it was the Patriots.*

People also ask me, "What was your big break in show business?" For me it didn't happen like that. Elvis Presley did one song on *The Ed Sullivan Show* and the country went crazy. I had no big break. It was a combination of a lot of things: sixteen *Ed Sullivan Shows,* seventy times on *The Tonight Show,* forty-five *Merv Griffin Shows,* twenty Lite Beer commercials, and owning a successful nightclub. Then twenty-five years ago, I did *Caddyshack,* which got me into movies.

Mine was not an easy road.

My image has its problems. People watch a guy degrade himself for an hour onstage, they get carried away and start to believe that it's really me up there. And with my image of "no respect," many people have treated me that way. One night I was doing a show at my club, and as I was about to walk onstage, a man sitting close by said, "Hey, Rodney, before you go on, do me a favor, will ya? Let me have your autograph—and more butter?"

One day I was taking a walk in Manhattan. A pretty girl recognized me and said hello. We got to talking, and she told me that she wanted to be a singer.

I didn't want her to have unrealistic dreams. "Show business is tough," I told her, "even for talented people. It's hard for anyone to make it."

And she said, "Well, *you* made it."

I guess she figured that if I'd made it, anybody could make it.

One day I was with some people in the coffee shop at the Riviera Hotel in Vegas. The waiter told me that a couple at another table said they'd buy me a drink if I'd sit with them.

I told the waiter, "Find out how long I have to sit with them to get a steak."

I was in another Vegas coffee shop having lunch with a few friends when I noticed that almost everyone in the place was staring at me. I was fairly popular then, but not *that* popular. I felt very ill at ease. I didn't know what to do. I decided to wave every once in a while. After I'd paid the check and got up to leave, I found out why everybody had been "staring at me"—I'd been sitting under a huge keno board.

I was in a bar the other day, having a few shots, and they told me to get out. They wanted to start the happy hour.

My father's funeral was one of the loneliest moments of my life. He was seventy-eight when he died.

I flew down to Florida for the funeral, and I was the only person there—my mother had died a few years earlier. Just as the guys with the ropes were lowering my father's casket into the grave, a guy came up to me and said, "Hey, Rodney, can I have your autograph?"

When I decided to get back into show business, all my

friends—and my father—told me I was nuts. He told me to get a regular job and forget show business. I'm glad that before he died he was able to see that I was successful.

A few years before, I was working in a nightclub called the Valley Stream Park Inn out on Long Island, and I asked my father to come along. This was way before I'd made it on *Sullivan* and *The Tonight Show*.

We walk into the club and see that they had a big party of four hundred people in the room. Only problem was that they were all Chinese. I said to my old man, "Look at this, will ya? All Chinese. What can I do here?" Of all nights to bring him, I pick this one. I knew I had a good act at that point, and I was really looking to impress my old man, but having four hundred Chinese people stare at me wasn't going to do it.

The nightclub manager could see that I was disappointed, so he said, "Don't worry. They are all college graduates and they all speak English. They'll understand everything you say."

He turned out to be right. They were a great audience and the show went great. I was so relieved that my father got to see me kill an audience.

Driving home, my old man said to me, "I think you've got something."

My ol' man was tough. He allowed no drinking in the house. I had two brothers who died of thirst.

It's not just me. Everyone—at one time or another—gets no respect. The movie *Fargo* got no respect. A few years back, it was up for Best Picture with *The English Patient.*

I saw *Fargo,* and like most people, I thought it was great. I saw *The English Patient* and thought it needed a doctor. Like a lot of people I spoke to, I didn't like the movie.

The Motion Picture Academy named *The English Patient* best movie of the year.

Now how could anyone who'd seen both those movies choose *The English Patient* over *Fargo?* I know how—it seems more sophisticated to like *The English Patient.* The name's so *classy*—whatever it means—and the story's so *serious*—whatever it was supposed to be about.

From day one, comedians got no respect from the Academy. Actors and actresses know that comedy is the toughest thing to do in show business . . . unless it's bird-calls. Many comedians are great actors, but few actors can do comedy. But comedy never gets respect. Laurel and Hardy, W. C. Fields, Mae West, the Marx Brothers, Jack Benny, Jackie Gleason, Bob Hope—none of these legends ever won a regular Academy Award.

Yet when the Academy has their awards show every year, they get a comedian to host it.

I told my landlord I want to live in a more expensive apartment. He raised my rent.

About ten years ago, AT&T was trying to lure back customers who'd switched to other phone services. Their big ad campaign said, "Call AT&T and we'll take you back." All you heard was "Call us, we'll take you back."

So I wrote a joke about it. "I get no respect. I called up AT&T. They won't take me back." I did it on *The Tonight Show,* and it got a good laugh.

That gave me an idea for an AT&T commercial, so I got in touch with some big shot there and told her my idea. She said, "I like it. Let me think about it for a week." A week later, she said, "Let's do it."

AT&T changed its mind four or five times over the next six months. I found out they took many surveys about whether people liked me, and asked them if they thought I would be good to do commercials for AT&T.

Finally, they called me and said, "Okay, it's definite. We're doing the commercial. We'll start right away."

But I had a problem. I couldn't start right away. I had to check into the hospital. I had an abdominal aortic aneurysm, which forced me to have a very distasteful operation. They cut my gut open, took all my intestines out, and put them on the table while they fixed my aorta. When they were finished, they stuffed all my intestines

RIGHT: One day, I was out on my balcony in L.A., and I saw my new neighbor, Shaquille O'Neal, on his balcony. We gave each other a wave. The next day, he sent his calling card: one of his size 22EEE shoes, inscribed To Rodney, I gets no respect—Shaq.

back in and stitched me up. For the next three or four months, I was in constant pain while my intestines shifted around, trying to settle.

I called the woman at AT&T and told her I had another "project" I was working on and had to finish it. I figured it would take about four weeks. She told me to call her as soon as I was free.

As soon as I was feeling myself again, I called AT&T and said, "Let's get going."

They flew a fellow from their New York advertising agency out to L.A., and he and I wrote two commercials in two days.

We liked them, AT&T liked them, we were set. I shot both commercials in one day.

A few weeks later, AT&T ran the commercials on TV, and they were a huge success. AT&T then got full-page ads with me in all the major newspapers, and radio, too. Almost immediately, the business started coming in. Everybody was happy, happy, happy.

They paid me well for the six months of that campaign, so I was planning to do some more. I had two more ideas ready, and the fellow from the ad agency was set to fly out to L.A. again to help write them.

Then out of the blue I got a call telling me there was some kind of problem. So I called the nice lady at AT&T, who told me that one of the company's big stockholders didn't want me in any more commercials because—he said—I wasn't *dignified* enough to represent AT&T.

Okay, I figured. *That's that.*

The other night I turned on my television and saw another comedian doing an AT&T commercial—someone I guess they think is more dignified—Carrot Top.

I tell ya, my wife was never nice to me.
On our first date, I asked her if I
could give her a good-night kiss
on the cheek. She bent over.

Chapter Ten
Let the Good Times Roll

I tell ya, my wife is never nice.
She won a trip to Las Vegas
for two. She went twice.

Thanks to my appearances on *The Tonight Show,* my career took an upswing.

I got booked for the first time in Las Vegas, opening for Dionne Warwick at the Sands Hotel. My salary: $4,000 a week for two weeks, which was a lot better than what I had been making selling paint and siding.

I had no trouble getting work now. In fact, I had more gigs being offered to me than I could handle, and I was on the road constantly. And it wasn't working for me.

I was now forty-seven and feeling old, so I decided to open a nightclub in New York, Dangerfield's. I opened it for one reason—I had to get off the road and be in New York to look after my two young kids. I wasn't living with my wife anymore, but her arthritis was so bad that it made it impossible for her to take care of our kids.

Twice in my life people told me I was nuts. The first

time was when I was forty and I decided to go back into show business. Everyone told me I was out of my mind. But I stuck to it and I'm glad I didn't listen to them. The second time was when my partner and I decided to open Dangerfield's. They gave me two weeks, a month. Thirty-five years later, we're still in business.

Opening Dangerfield's wasn't easy. At that time, I had maybe $50,000 in the bank, so I went to all my friends borrowing money—$2,500 here, $5,000 there, another $2,500 over there somewhere. It was hard—and scary—but it made me feel good that so many of my friends had that kind of confidence in me.

The original cocktail napkin from my club. We still use them today.

I loaned a guy $10,000 to get plastic surgery. Now I can't find him. I don't know what he looks like.

I was told I would need $125,000 to open the club. It ended up costing me $250,000. It was supposed to be ready in two months; it took about six months. We opened the doors on September 29, 1969.

While my partner, Anthony Bevacqua (affectionately known as "Babe" because he was the youngest in his family), was overseeing construction of the club, I was working in Vegas. And every night after my show, I'd run all over town getting my picture taken with all the big stars playing there. We then put all those pictures on the wall at Dangerfield's. We've got dozens of them.

We had a practice run the night before we opened— invited about a hundred people to see the show and eat and drink for free. I knew we were ready, but I was still as nervous as hell. I remember sitting in that empty club with Babe a few hours before we opened the doors for the first time, frantically trying to think of something we might have forgotten. After a few minutes I thought of something— we had forgotten salt-and-pepper shakers for the tables.

Other than that, everything went great on our preview night.

We opened for real the next night, and never looked back. We were off and running. We drew young people and old people, tourists and natives. We were even big with the rock-and-roll crowd. I can't remember all of them—guys with long hair all look alike, you know?—but I know that the guys in Kiss and Led Zeppelin used to come by whenever they were in town. And after about five or six months, we were doing so well that I paid back all the money I had borrowed from my friends.

A month later my accountant told me, "You have no money left, and you have to pay taxes. All that money you paid back to your friends was supposed to go to the IRS."

So now I had to go back to all my friends and borrow the same money *again.*

After that screwup, I saw no profit from the club for about two years. That's how long it took us to pay back all my friends again, to pay the IRS, and to pay off the ice machine. This put me in a weird position, because I was now getting really hot in the business, but I couldn't take jobs out of town because I had an obligation to my partner, and to my young kids again. I had to turn down good jobs, much more money than I could make working at the club.

Oh, the other night my wife met me at the front door. She was wearing a sexy negligee. The only trouble is, she was coming home.

Before I opened Dangerfield's, people warned me, "If you start doing good business there, the Mafia will take it from you." But I wasn't worried about that—I'd met a lot of Mob guys when I worked at the Copacabana, and I got along great with them. When I opened Dangerfield's a lot of those guys came in, wished me good luck, and were my best customers. I never had a problem with them.

Some of those Mob guys were funny. One night this "made" guy was talking about the nicknames he and his friends had given one another when they were kids. He mentioned one nickname a few times that I never forgot. He referred to one kid as Mile-away.

I said, "Why'd you call him that?"

He said, "Because whenever there was trouble, he was always a mile away."

The toughest club I worked was owned by a guy named Nunzio. Man, he was tough. One day he said to me, "Kid, you wanna go hunting?" I said, "Okay, I'm game." And he shot me.

Once when I was backstage just as Ed Sullivan was about to introduce me, I could hear a couple of the stagehands talking about my club Dangerfield's. Sullivan says, "And here he is . . ." and one of the stagehands yells to me, "Hey, Rodney! Can I get laid at your joint?"

As I walked onstage, I yelled back at him, "Leave my joint out of this!"

I bought another book, How to Make It Big.
I got ripped off. It was about money.

They say an elephant never forgets. Around this time, I had a night with an elephant that *I'll* never forget.

One night a woman came to see me at Dangerfield's. She asked me to do something for charity. They wanted celebrities to perform with the circus, to ride around Madison Square Garden on elephants.

I said, "Sure," and two days later, I report to the circus at Madison Square Garden, all set to ride the elephant. The trainers bring the elephant out, and they hoist me up into the saddle. No problem. Now we start walking.

We're maybe halfway around the ring, and I find myself having a problem staying on the elephant. He's moving quite a bit from side to side—he's swaying and I'm slipping. Sure enough, he swayed this way, I swayed that way, and next thing I knew, I swayed my butt right off the elephant.

RIGHT: Celebrating our tenth year at Dangerfield's, with my partner, Tony Bevacqua.

Being thrown off an elephant was bad enough, but my ankle got stuck in the stirrup, so the elephant was dragging me around.

I was scared. "Hey, help me!" I'm yelling. "I'm stuck!"

But nobody took me seriously. People in the crowd were laughing, waving. "Hey, Rodney, how ya doin'?" They thought it was part of the act.

When the elephant was walking, every time his rear left foot came down, I had to make sure it missed my head.

Finally, two guys who worked for the circus saw it was serious. They stopped the elephant and got me untangled. When I got to my feet and had smacked all the hay off me, I gave the elephant a dirty look.

Then I thought, *Ah, forget it. He don't know what he's doing. He's a dumb animal.*

To show the elephant I wasn't mad at him, I started feeding him peanuts. Two minutes later he left me for a guy who had cashews.

---

*I tell ya, I know I'm ugly. My dog closes
his eyes before he humps my leg.*

I got another big break in 1970, when I did my first Lite beer commercial. That was a good gig for me—I went on to do fifteen or twenty more. I mostly hung out with

Bob Uecker, Deacon Jones, and Bubba Smith—all funny guys. And they were former professional athletes, so there was nothing I could teach them about having a good time.

There were a few things I could stand to learn, though. We were shooting one of those commercials on a beach in Florida and I noticed that all the girls were *amazingly* beautiful and the guys were all so handsome. I turned to the guy next to me and said, "Wow, have you ever seen a beach like this? No fat people. Only young beautiful people. This is the place to hang out. Maybe I should buy a condo down here."

This guy looked at me like I was an idiot and said, "They're all actors. They were hired for the shoot."

I remember my first meeting with the famous Canadian hockey star Boom-Boom Geoffrion when we did a beer commercial together. When they introduced me to him, they said, "Rodney, say hello to Boom-Boom."

I said, "Hey, Boom-Boom, I know your sister, Bang-Bang."

I was doing a baseball bit for another beer commercial, and Bob Uecker had to throw the ball between my legs. He throws the ball pretty hard. I was worried that he might hit me with it.

So I said to him, "Be careful."

He said, "Don't worry. I'll throw it around your knees."

I said, "That's no good. You'll hit my cock."

When I go to a nude beach, I always take a
ruler, just in case I have to prove something.

One night after the show at Dangerfield's, I mentioned to some people at the bar that I was having a bad night, that I was really feeling down. I got plenty of advice on how to get rid of my depression.

Of course, I got the usual "it's all in your mind," which was the stupidest thing I'd ever heard. That's like telling an ugly girl, "It's all in your face . . ."

A girl said, "Hire a limousine and have the guy drive you around Manhattan. After half an hour, you'll have no more depression."

I said, "Don't tell me that the forty-eight Austrian psychiatrists I've seen, all the money I've paid them, their advice meant nothing, that all I got to do is ride in a limousine, and I'll be cured?"

A Mob guy gave me his solution for depression. "Come with me," he said. "We'll go to Vegas. We'll fuck whores. You'll feel better."

I didn't hire the limo . . . or the hookers. But I got a good laugh at the advice, which made me feel better.

RIGHT: You're a great crowd. That's it folks, show's over.
I'm going backstage now and take a shower. Maybe I'll
get lucky with the soap.

In my life, I've talked with many psychologists and psychiatrists. It has cost me a lot of money, but at least I got a few jokes out of it. I think the first one was: "I told my psychiatrist, 'I keep thinking about suicide.' He told me from now on I have to pay in advance."

One day I was on my way to see my psychiatrist, but I had to make a deposit at my bank. While I was standing in line, the bank guard started talking to me. After a few minutes, I said, "Hey, I gotta run. I'm late to see my psychiatrist."

He looked at me kind of puzzled and said, "You need a psychiatrist? A husky guy like you?"

---

*I told my psychiatrist, "Doc, I keep thinking
I'm a dog." He told me to get off his couch.*

One night I was in my dressing room at Dangerfield's before the show, and the maître d' told me that Johnny Carson was on the phone.

Carson said he wanted to come down, see the show, and asked if I had room. I told him, "Johnny, it's Saturday night, first show—we're full. But for you, whatever you want, as many as you want, you got it. How many will there be?"

*RIGHT: Hanging at Dangerfield's with that wild and
crazy guy, Steve Martin.*

He said, "Just me."

I said, "Come on down, Johnny. No problem."

Back then, I had a buddy, Dave Goldes, who had worked for Johnny on *The Tonight Show*. Dave was an original, funny guy, and brilliant—a Rhodes scholar and a poet—but quite weird. He's an excellent comedy writer, and he wrote several jokes for me. One I still use in my act: "I feel sorry for short people. When it rains, they're the last ones to know about it."

A couple years before, I got Dave a job writing for *The Tonight Show*. Carson liked Dave's jokes but had a little trouble with his personality. Dave was always depressed, always down. He was not sociable. He wouldn't sit with the other *Tonight Show* writers at the big table. He'd sit off to the side. And he wouldn't dress up for anybody. I liked him, but let's face it—he was a weirdo. So after a while, they let him go.

At the time I said to Johnny, "Who cares if he's not social and he doesn't dress right? He brings in the jokes, right? Who cares if his socks don't match?" So they hired him back, but a few months later they let him go again.

Despite this history, I knew Johnny liked Dave, so I called Dave and told him Carson was coming down in case he wanted to drop by and say hello.

Twenty minutes later, Carson's in my dressing room. We're sitting there talking, and I said to Johnny, "Wanna drink?"

RIGHT: With a young Robin Williams, back in his Mork days. What I went through to get him to stand still for this picture.

He said, "No, no."

"I know how it is," I said. "I like to have a drink . . . or two . . . too. So if you feel like having a few drinks, I'll make sure you get home all right."

Carson smiled and said, "Okay, give me a double Scotch."

Pretty soon he was feeling good, and the show was about to start, so I took him upstairs. We put a small table off to the side, where no one would bother him, and I go up to do my act.

After a while Dave showed up. I guess he was just trying to be funny, but his clothes for the evening were a burlap bag and a pair of sandals. The club was jammed, so the maître d' put Dave at the same table with Johnny.

When I finished my act, I joined them. I sat down, and Carson called the waiter over. "Let me have another Scotch," he said, then looked at Dave. "And a pair of socks for my friend."

You wanna really confuse a guy?
Join him while he's taking a leak in the street.

A couple of years later, the Dangerfield's maître d' called me in my dressing room. He said Jack Benny was on the phone from L.A. When I pick up the phone, Jack said he'd seen me on *The Tonight Show* that night.

"Rodney," he said, "when I'm watching someone three thousand miles away and they make me laugh, I have to call them. There was one joke you told that would have been perfect for me. You were talking about your wife's cooking. You said, 'And the way she serves a meal. You put down a steak. How do you forget the plate?'"

I told him he could use the joke, but he said, "No, I wouldn't do that."

He was right, though. The joke was better for him.

One of my most memorable nights at Dangerfield's was when Jack Benny came in. After I did my act, I joined him at his table. He was very complimentary, which was, in my mind, like getting praise from God.

Benny was class, a real gentleman. After we'd talked for a few minutes, he said he and his friends were going to a nearby restaurant to get a bite to eat, and asked if I'd like to join them.

I told him, "Gee, I'd love to, but I'm writing something that I have to finish tonight."

He said, "I understand. That's okay."

The truth was that I didn't go because I knew I couldn't be myself with Jack Benny. I mean, I'd have to play a part and be a gentleman. Can you picture me saying to Jack Benny, "Man, I'm so depressed. It's all too fucking much"?

My wife can't cook at all. She made chocolate mousse. An antler got stuck in my throat.

Chapter **Eleven**
A Night with Lenny Bruce

*With sex, my wife thinks twice before
she turns me down. Yeah, once in
the morning and once at night.*

anging out with Jack Benny and Johnny
Carson—you can't do any better than that. But
my first brush with fame was back in the early
forties, when I did a show with Al Jolson.

I was working at a nightclub in Atlantic City called the
Paddock. One night the boss told me that Jolson was in
town doing a benefit show and needed a couple of acts to
go on before him.

I said, "Sure." Anything to be on a stage with Jolson.

A few hours later, I was backstage in a theater looking
at Jolson standing in his underwear, reading a telegram.
He was disappointed. He said, "Why couldn't it be from a
girl?" It's been rumored that Jolson liked to have sex
before he did a show. Apparently that night he struck out.

Even stars can't get lucky every night.

*The day my wife and I got married—
that was a beauty. I gave her the ring
and she gave me the finger.*

A few years later, I was working at a nightclub called
the Queen's Terrace on Long Island. One night Jackie
Gleason walked in. He sat down at a table all by himself
and watched the show while he drank a whole fifth of
Scotch. Then he got up, gave the waiter a $20 tip, and
walked out, straight as an arrow.

Gleason had quite a reputation as a drinker, and he
liked to enjoy himself. He proved that night that he could
do both. Jackie had an appetite for other pleasures as well.

One night I was in Bobby Byron's room at the
Belvedere Hotel with Joe E. Ross. There was a knock on
the door.

Bobby yelled, "Who is it?"

I heard a voice say, "The Great One."

Gleason called himself "The Great One," and he was
worthy of the title.

Bobby opened the door and did a little business trans-
action—he sold the Great One some great pot. And away
we go . . .

RIGHT: I went fox hunting . . . look what I found.

Hey, I gotta be honest with you. I'm not a
fabulous lover. My wife and I were in bed on
our wedding night and she said, "Well, honey,
this is it." I said, "Honey, that was it."

I first met Lenny Bruce through his mother, Sally Marr, who was the mistress of ceremonies at the Polish Falcon Club, where I worked as a singing waiter back in 1942. As the years went on, I became friendly with Lenny—we saw each other here and there and hung out. This was years before he had an act, or even knew he was going to have an act. Lenny was the nicest kid in the world, and he idolized Joe Ancis, one of my oldest and dearest friends. I'll tell you a lot more about Joe later.

Lenny hit it big in the fifties. One time I was hanging out with him in his room at the American Hotel in New York. He was getting ready to do his act at a club in the Village, but first he had to shoot himself up. Even then—young, smart, and at the top of his game—he had a bad heroin habit.

It was difficult for me to watch as he held that needle, looking for a good vein in his arm, so I went into the bathroom until he had finished. When he was done, he was a new man. He said, "Tonight I'm going to do the show dressed in white—completely white, everything."

After he became the man in white—white pants, white

shirt, white jacket, white shoes, white socks, white underwear, far as I can remember—we went down to the club.

Here are the first words Lenny said that night when he took the stage: "Tonight, here's how I'm going to open my act. I'm going to pee on you. If a guy tells jokes, you'll forget him. But if a guy pees on you, you'll never forget him."

A guy in the audience yelled, "Keep it clean. Keep it clean."

And Lenny answered him in these exact words: "Fuck you, Jim, you square motherfucker!"

We were poor, too. If I wasn't born a boy,
I would have nothing to play with.

Redd Foxx was another good friend of mine. He was a great comedian, a "natural," as they say. (I've learned over the years that when you get to know the "naturals," you find that they work very hard at their craft.) Here's one of my favorite Redd Foxx jokes: "China has over a billion people. Let's face it—they outfucked us."

One time, when I was very tense and had been working at my club nonstop, I needed to get away, so I called around to find out where Redd was performing. He was working in Florida, so I flew down to see him. My first night there, I put so much stuff up my nose that when I went to sleep at eight in the morning, I said to myself, *If I wake up, it's a gift.*

Later, I was doing a TV special for ABC, and I wanted Redd to be in it. He was working in Las Vegas, so we had to fit it into his schedule. In show business, if you're not working one night, they say you're "dark" that night, so I said, "Redd, are you dark on Monday night?"

Redd said, "Rodney, I'm dark every night."

In the 1980s, I did three comedy specials for ABC in L.A. We did one of them during the summer, so I rented a house in Malibu.

My next-door neighbor was Flip Wilson. We would sometimes hang out at night and swap stories. Flip told me a disturbing story from when he was an unknown working in the South.

Flip had just closed in one city, and a man he'd met at his hotel told him that another guest was driving to the same city Flip was going to next, so Flip asked the guy for a ride.

The guy said okay, so Flip got in the car and off they went. After they'd been driving for a while, the guy said, "What kind of work do you do?"

Flip told him, "I'm an entertainer."

The guy said, "All niggers are entertainers. What do you do for a living?"

I first saw Andy Kauffman at a club in New York called Catch a Rising Star. What amazed me was that Andy did two shows each night, each one completely different. For the second show, he'd come out as this obnoxious character, and the audience never knew it was Andy doing both spots.

I immediately became a huge Andy Kauffman fan and asked him if he'd like to open for me on a couple of dates. He said okay, so we went to work at a theater in San Francisco.

Andy made life interesting, on and off the stage. Onstage, his thing was to get the audience to hate him. He did an amazing character, a repulsive singer named Tony Clifton, who told vile jokes and said nasty things about women and insulted everybody. And the audience hated him—they'd curse him out and even throw things at him. And that's exactly what happened on our opening night in San Francisco.

They threw so much stuff at Andy, when I came on it looked like I walked into a salad bar.

The next night, the theater manager was prepared—he put up a net that covered the entire length of the stage to protect Andy from people throwing things. It worked pretty well—only two things got through the net and hit Andy: an apple and a hard-boiled egg.

I loved it, and Andy loved it. He was behind the net with thousands of people yelling at him, "Get off! You stink! Get off, you bum!"

Andy, still in character as Tony Clifton, raised his hand to get the audience to be quiet and said, "There are a few of you out there who are ruining it for the rest of us."

People said to me, "What are you doing with him as

your opening act? You need someone who puts people in a good mood. They hate this guy."

I said, "No, he's different. I like him."

My second date with Andy was at the Comedy Store in Los Angeles. Before the show, Andy had the emcee make this announcement: "Ladies and gentlemen, Tony Clifton asks that there be no smoking during his performance." There was some grumbling—remember this was thirty years ago—but people put out their cigarettes.

About five minutes later, just as the show was about to start, the announcer once again said, "Ladies and gentlemen, please remember that when Tony Clifton is on, there can be absolutely no smoking during the show."

Then he introduced Andy as Tony Clifton—and Tony walked out smoking a cigarette.

---

*You know you're drunk when you take
a leak and your fly isn't open.*

In 1985, HBO offered me a deal to do a show from Dangerfield's that would introduce new comedians. They told me I could pick the comedians.

I liked that idea because I've always been looking out for new talent. I was always attracted to the edgy comics—like Andy Kauffman and Sam Kinison.

I did an HBO show every year for six or seven years, and they were a big hit with viewers, and with comics. Most of the young comedians I contacted wanted to be on my shows. Many times, it was their first shot on national television, and it led to bigger things for a number of them. Here are some of the people I introduced on those shows: Jerry Seinfeld, Robert Townsend, Roseanne, Bob Saget, Louie Anderson, Andrew Dice Clay, Carol Liefer, Rita Rudner, Jeff Foxworthy, Tim Allen, and Sam Kinison.

He who laughs last didn't get
the joke in the first place.

About Fifteen years ago I was working at Caesars Palace in Vegas. My opening act was a girl named Pam Madison, who sang and did impersonations of singers. I'd never worked with her, so before our first show, I asked her what her closing number was. The opening act usually does about a half hour, but I wanted to know a little more precisely when I had to be ready to go on. She told me she always closed as Barbra Streisand singing "People." Okay, fine.

So, opening night, I'm in my dressing room, just pulling myself together as Pam goes on for the first show at eight. I hear her introduced, I hear the applause, whatever . . . I

stop paying attention. The next thing I know, I hear Pam doing Barbra Streisand singing "People"! I'm confused as hell, so I look at my watch. It's only eight-twelve! She'd only been on for twelve minutes! So I frantically tie my tie, put my shoes on, and run toward the stage.

After the show I asked Pam what had happened. She told me, "I saw everyone in that audience looking at me and I panicked. I had to get out of there."

She settled down by the next show, and we got along great. One night after the show, Pam introduced me to a friend of hers who was working at a comedy club next door. It was Roseanne Barr, who was an unknown at the time. The three of us sat and talked for a while and I could see that Roseanne was very funny. I also thought, *What a voice! She'd be the perfect wife to abuse me.*

Before we parted, I told Roseanne that I was doing one of my HBO comedy shows in a few months and I'd like her to be on it. I told her that I wanted her to do her stand-up act, and that I'd also write some skits that she and I could tape for the show. She said, "Sure, whatever," but didn't seem too excited.

Three months later, I called her and said it was all set—skits were written and I was sending her a copy of the script for her to look over. Roseanne now got *real* excited and told me, "This is the first time someone told

LEFT: Bob Hope entertained the troops every year. So did my wife.

me they were going to do something for me and they actually did it!"

As you might have guessed, Roseanne was a hit, even though it was a big show—I also had Jerry Seinfeld, Sam Kinison, Robert Townsend, Bob Nelson, and Jeff Altman on that night.

Roseanne and I stayed friendly. Years later, in the early nineties, we hooked up at the Pritikin Longevity Center in Santa Monica, which was where people used to go to lose weight and get healthy. We spent a week there with Carl Reiner and his wife, Estelle. We all shared the same table for our meals and we had plenty of laughs, but one night Roseanne and I were having such a good time, laughing so hard about the awful—HEALTHY—food they were serving us, that we both flipped and decided to escape for a night. After we finished our dinner at Pritikin, which was not very tasty, we decided to go out and *really* eat. We both went nuts and ate all the junk we'd been craving all week.

My wife can't cook at all. I got the only dog that begs for Alka-Seltzer.

I met Mike Tyson a few times, at the clubs and in Vegas, and he was always very nice to me. He's not as scary as he looks, at least to me. Maybe that's because I made him smile the first time we met. I looked him in the eye, real concerned, and said, "If anybody bothers you, let me know."

I live in a bad neighborhood. Last week they raffled off a police car, with two cops still in it.

I once had a press agent named Lee Solters, who thought he was a big shot. He'd been Frank Sinatra's publicist for thirty years, so maybe he was. Anyway, I noticed that whenever I talked to him on the phone the conversation always ended the same way. After a few minutes he would say, "I gotta hang up. I got a call coming in from overseas." The first few times I believed him. Then I realized that this was his way of ending the conversation . . . and to make himself look like a big shot. After that, I would end our conversations by saying, "I gotta hang up. I got a local call coming in."

What good is being the best if it brings out the worst in you?
—LADYBUGS, *1992*

In L.A., about fifteen years ago, I ran into a guy named Ron Jeremy, a porn star who grew up in New York. He was a smart guy, funny, and I took a liking to him. Even put him in one of my HBO skits. One day he said, "You know, Rodney, I really want to move back to New York. The people in L.A. are too crazy."

Now, Ron Jeremy first became famous for a movie in which he said to a girl, "How about a blow job?" and she blew him off by saying, "Why don't you give *yourself* a blow job?" He said, "Not a bad idea," and went down on himself.

It was a wild thing, and people talked about it. That's how he became famous.

That's why I broke up when he said he wanted to move to New York—a guy who makes a living going down on himself thinks people in L.A. are too crazy.

---

*Last week, I went to a discount
massage parlor—it was self-service.*

Speaking of big movie stars, I first met Dustin Hoffman at my club back in the early seventies. He was about to play Lenny Bruce in Bob Fosse's movie and he wanted to get the feel of a nightclub and of doing stand-up. So he came over to Dangerfield's and hung out for a week—

although I told him, "You can't learn to do stand-up in a week. You have to walk the boards for twenty years."

One night Joe Ancis, who was Lenny's longtime friend and mentor, dropped by to say hello. In the course of our conversation, Dustin said to Joe, "Do you think I'll make a good Lenny Bruce?"

Joe said, "No."

Dustin wasn't exactly thrilled to hear that, but I told him, "Don't worry about it. Very few people ever saw Lenny work, so no one will know the difference. Whatever you do, they'll figure *that's* Lenny."

Dustin went on to win an Oscar nomination for *Lenny*.

One of the most talented guys I ever saw was Sam Kinison. I first met him in the early eighties when I was working in a place called the Arena, a theater-in-the-round in Houston. After the show, a few of us went to a local comedy club.

The show was Sam and two other young comics. He was young, raw, but had something wild that I liked. After the show, I asked the club manager, "Can I meet the guy who went on second?"

They brought Sam over to our table, and we talked for about fifteen minutes. He told me he was struggling with his act, but I told him, "You'll be fine. You're great."

Sam had a crazy background—he'd been a preacher and came from a family of preachers. So there was plenty of anger and turmoil in him, which came out brilliantly onstage. Hilarious—but full of obscenities and rough talk

about sex and drugs and religion. Sam didn't hold back. He was raw, and honest, and very funny.

I met him again a few years later in New York at Catch a Rising Star. After we said our hellos, Sam said, "Rodney, I finally got it, man!"

So I stayed for his act, and he was right. He really did have something. His material was still wild, but his delivery was now mature. He was sensational. Sam, who'd been divorced three times, said he wasn't going to get married again—"I don't have to give away everything I own *every five years!*"

After that show, I said to him, "Jeez, you're great, man. I'm doing an HBO show. I'll put you on it, okay?"

He said, "No. That format's no good for me. I can't do just six or eight minutes like that. I need more time. I can't cook until I'm out there at least ten minutes."

I said, "You're wrong, man. You can cook as soon as you walk onstage. Just go right into it."

"I don't know . . ." he said.

Sam went back to California, and two weeks later I got a phone call. It was Sam, and he was very excited. "Rodney," he said, "I'm cookin' right away now! Can I still get on your show?"

I said, "You got it."

So he came out to New York and did my show, and he was hilarious. That's when he did his now-famous Ethiopia

LEFT: Here's Sam Kinison giving me a lecture on how to handle women.

joke. The papers were full of stories about a horrible famine in Ethiopia, so Sam said: "Hey, I've figured out why you people are starving—you live in a fuckin' desert! See this? This is sand. NOTHING WILL GROW HERE!"

After that show aired, Sam started packing places that seat four or five hundred people. He used to say that that HBO show was "the six minutes that changed my life."

About a year later, I said to Sam, "I'm doing another special. Kill 'em on this one and you'll be playing two-thousand seaters."

But I was wrong.

He did the show and killed 'em again. The next time I saw Sam he was headlining at the Universal Amphitheater in Los Angeles. *Six thousand* seats—packed—and everyone yelling, "Sam! Sam! Sam!"

Sam once told some writer that "Rodney is like a god to me." That touched me deeply . . . but he probably said it because he thought I'd saved his life, not his career. Let me explain.

Before that second HBO show, Sam and his brother Bill came to see me. They were very worried. They said Sam had pissed off some Mafia guy who was now after him. They thought this guy was going to kill Sam.

I said, "Ah, I don't know about that . . ." I'd met plenty of Mafia guys, and I was pretty sure Sam had nothing to fear. So I told him, "You got nothing to worry about. Concentrate on your show. The guy's just trying to scare you. I'm telling you, nothing's going to happen."

Sam said okay, and he felt better. And of course, nothing happened to him. But because I had been so sure that nothing was going to happen, Sam and his brother got the idea that I had made a phone call to someone in the Mafia. Sam and Bill were now under the impression that I was connected, that I was "mobbed up."

If you never saw Sam's act, do yourself a favor and buy one of his albums, or all of them. Sam had a rather strange sense of humor. When his father died, Sam's mother was very morose. To console her, Sam said, "Ma, look at it this way. You'll have more closet space."

Sam had a lot of great bits. Here's one: "Today, everybody's sick. They even have dog psychiatrists. I'd like to get in on that racket. 'What's that, Mr. Raven? I'm sure Spot's a good boy. Let me take him inside and talk to him for a few minutes. Wait here. I'll see what I can do.'" Then Sam would walk to the other end of the stage and pretend to open and close a door. Then he would yell at the imaginary dog, "What the fuck are you doing? Behave yourself! You're a fucking dog! You shit in the street."

When Sam got hot, I told him to make sure he only worked in places where he could be himself, where he didn't have to censor himself. "You're a big hit on HBO," I told him. "But if you go on network television, they'll be cutting you and cutting you. 'You can't say this, and you can't say that.' It'll drive you crazy."

So I suggested that he stay with HBO. I said, "The audience will find you."

About six months later, Sam's manager convinced him to do *Saturday Night Live*. He said, "It's *Saturday Night Live*! How can you turn it down?"

So Sam did it, and he had huge problems with the censors for doing two bits that the network guys had told him not to do. One was about the drug war and the other was about Jesus getting nailed to the cross.

Since the show was live from New York, people on the East Coast saw Sam's full performance, but his bits were cut for the rest of the country.

Sam wasn't just banned from *Saturday Night Live,* he was banned from NBC, which meant no *Tonight Show*.

That would have killed the career of a lot of comedians, but it didn't hurt Sam. He just kept rolling.

If you saw my movie *Back to School,* you'll remember Sam as the crazed Vietnam vet–history professor. As soon as I knew the movie was a go, I wrote a part for Sam. He exploded on-screen, as I knew he would.

Sam was killed in a car accident while driving from L.A. to Vegas in 1992. He was just thirty-eight.

I went out to L.A. for the service, to pay my respects. He was lying there in a coffin and it was so strange, because the morticians had made him look so peaceful. He seemed to be smiling—like he was happy to be in that damn box.

I looked down at him, lying there, and just to make sure, I, like an idiot, whispered, "Sam . . . ? Sam . . . ?"

He didn't move.

I thought to myself, *He's either dead—or very snobbish.*

Thinking about Sam dying and movin' on reminded me of one of his very best bits:

Up in heaven they asked Jesus to come back down to earth. They said, "C'mon, Jesus, it's been two thousand years. Why not go back down, spread a little peace and joy." And Jesus says, "Sure, man . . . JUST AS SOON AS I CAN PLAY THE FUCKIN' PIANO AGAIN!"

I first met Tim Allen when he came to New York from Detroit to audition for one of my HBO specials. He was very funny and I wanted to put him on the show, but I screwed up—I had too many guys I was obligated to use for that show. When I told Tim, he took the news like a gentleman. When we did the next HBO show, he was my first call.

Years later, he was a big star, with his own sitcom, *Home Improvement.*

Tim is not just a guy who tells jokes. He proved himself to be a very good actor.

When I did a guest shot on Tim's show, I felt funny talking money with Tim's people, so I said to the show's producers, "I tell ya what. I don't want to hassle with lawyers and things. Tim knows me, so pay me whatever you think is fair."

Okay. Good. Everybody's smiling.

I do the show, everything is great. But when they gave me my paycheck, it was much lower than what I had expected. I didn't think they had been fair, but I forgot about it.

A few months later, Disney—which owned ABC and Tim's show—came to see me about a project. We did our business, and then I said, "Let me ask you something. Do you think what you paid me for *Home Improvement* was fair?"

"There's nothing we can do about it now," they said. "What's done is done."

"I tell ya what," I said. "The raincoat I wore on the show was nice." Some shows will give you your wardrobe, but not Disney. They keep it all. I said, "Leave the money where it is, throw in the raincoat, and we'll call it even, okay?"

They said, "Fine."

A week later, they sent it to me. It turned out to be an Armani.

Now I'm the only flasher with a $5,000 raincoat.

My wife has a temper. She keeps yelling
at me, "You're an animal, an animal!"
So I took a leak in the living room and
I told her from now on that's my territory.

---

first met Jim Carrey when I was working in Toronto, back in the early eighties. Someone told me to catch a young local comic who had something special. They were right.

Jim was about nineteen or twenty then, which means we've now been friends for more than twenty years.

Anyway, I caught his act in Toronto, and I thought he was fantastic. His mimicry was wild, and he had a rare gift for physical comedy. I saw he had talent to burn, and when he smiled you had to love him. So I hired him to open for me on a few dates.

He did a few one-nighters with me in Canada, and then I took him to Las Vegas to open at Caesars Palace.

When Jim first started working with me, he was mainly an impressionist, a sensational one. Who else can take his face and make it look like Mao Tse-tung, Brezhnev, Clint Eastwood, and Sammy Davis, Jr.?

But Jim wanted to stop doing impressions and be himself onstage. I could relate to that, because I'd gone through the same kind of thing as a young comedian. I used to do an impression of Al Jolson singing "Rock-a-bye Your Baby." It was good enough that I would close my act with it, but I got to the point where I had to lose it, because I wanted to create my own voice, my own identity onstage.

So I understood what Jim was trying to do with his act, and I knew that it wouldn't be easy. When he stopped doing the impressions and just did his bits, it didn't go over with the audience. He was bombing every night.

Most headliners would say, "I gotta get a guy who gets laughs." But I just said, "You're great, Jim. Don't worry about it." I knew Jim had something special.

In 1994 the American Comedy Awards honored me by giving me the Creative Achievement Award. Jim Carrey was very gracious when he accepted the task of introducing me and presenting me with the award. I was very moved by his words. His introduction of me was one of the highlights of my life.

I live in a rough neighborhood. One night I was held up, but the guy had class. He used an electric razor. This guy, he took my watch, my wallet, and a little off the sides. Actually, I blame myself. I was standing right next to an outlet.

People often ask me what it was like to meet Elvis Presley. While I was working at the Sands in Las Vegas back in the late sixties, I got an invitation to attend Elvis's closing-night party at the Hilton Hotel. So when my show was over, I went to the party.

Elvis was very warm and friendly. He walked in and said, "Hey, man."

I said, "Hey, man."

With our "Hey, mans" out of the way, we had somebody take a picture of us, and we chatted awhile.

LEFT: *I have no respect left. I gave it to Jim.*

I still have that picture on the wall at my club, which is why Dangerfield's has the distinction of being the only nightclub robbed over two thousand times. When Dangerfield's opened, we put that picture of Elvis and me on the wall in the bar area, along with the many other pictures of me with big shots. That picture has been stolen at least once a week—more often during prom season. In fact, the club has been open over thirty years and that's the *only* picture that has been stolen.

One night I was standing in the bar area and a girl said, "Hey, Rodney, can I take a picture?" She was kind of cute, so I straightened my hair and my tie and said, "Sure, honey, go ahead." She said, "Thanks," grabbed the Elvis picture off the wall, and walked out.

I was working at the Tropicana Hotel in Vegas in the seventies when those Elvis impersonators became so popular. During that period there were ten Elvis impersonators working on the Strip at the same time. You didn't even have to look like Elvis—if you wore an Elvis outfit and sang an Elvis song, you were a big hit.

One night I went backstage to visit an Elvis impersonator before his show. This guy was fat and very unattractive, but I heard him tell somebody, "The girl I want you to bring back after the show is the third girl I give a rose to."

RIGHT: This photo's been stolen hundreds of times from my club. I told Elvis it was a real kick to meet him. So many people tell me we look alike.

Then he grabbed his guitar and waddled out to a screaming audience.

Years ago, when you spent $25 to see a show, you saw
a star. Today, you pay $50 to see someone impersonate a
star.

---

*My wife and I, we both love Las Vegas. She likes
to play the slots, and I like to play the sluts.*

I met Barbra Streisand when I was tapped. It was at Elvis
Presley's party in Vegas. I was sitting at a table talking
to a couple of guys. All of a sudden I felt a tap on my
shoulder. I look up and it's Barbra Streisand.

She was very nice. She said, "Rodney, I just came over
to say good night." She was about to leave.

I said, "Barbra, remember the Bonsoir in Greenwich
Village? That was the first time I saw you." I recalled the
story to her.

I was there with a friend of mine. We were ready to go
home when the maître d' said to us, "There's a girl auditioning now, and I hear she's very good. You might want
to see her."

I said to my friend, "I tell you what. We'll get our
coats, we'll stay for one song, and we'll cut."

She came on, and about forty minutes later we were still standing there holding our coats in our arms. We were mesmerized by the voice and the whole thing. That's the first time I saw Barbra.

The next time, they're having a show in New York—a benefit show in Barbra's honor—and they asked me to be on the dais. So I do about four or five minutes. One of the jokes I told was about that night:

The first time I saw Barbra at the Bonsoir, it was a very unusual evening for me. It was the first time I went over the minimum.

I met President Reagan when I performed for him at the Ford Theater in Washington. Prior to the show, we all went to a reception at the White House, where I had my picture taken with both the president and Nancy Reagan.

After the show, I didn't feel like hanging out with the senators, governors, and all the other political big shots— I got high on some shit and went outside and hung out with the limo drivers. I figured I had less chance of getting in trouble out there.

The next day I showed the picture of me and the Reagans to my housekeeper Thelma. I said, "Hey, do you know this couple right here?" She looked at the picture. Her eyes got real big, and she said to me, "Did you meet them in person?"

My wife is the worst cook in the world.
At my house, we pray after we eat.

President Reagan wasn't the only president I met. I like to tell people that President Clinton slept on my floor. In the nineties, I was living at the Beverly Hilton Hotel when Clinton came to L.A. for a speech. He decided to stay in my hotel that night, so they gave him a suite on the eighth floor, the same floor I was on.

The Secret Service said that no other guests could stay on the eighth floor because the president was sleeping there. They had moved all the other guests out, and were getting ready to evict me, but Clinton said, "Rodney doesn't have to move. He's okay."

I thanked him, and we had our picture taken together. He was very cordial—we chatted about fifteen minutes. When I got back to my rooms, though, the Secret Service was there. They had two dogs running around, sniffing everything. I had some pot in my place, so I was sweating it out.

I got away with it. And I tell ya, these dogs, they were hip. On the way out one of them winked at me.

LEFT: *I was hoping to show the president how*
to inhale, but I never got the chance.

I tell ya, the dog drives me nuts. Last
night he went on the paper four times . . .
three times while I was reading it.

---

did *The Tonight Show* one time when Bill Gates was on.
Before the show, he told me he was a little nervous.

I said, "Bill, what are *you* worried about? You just
have to talk. I have to get laughs."

I wanted to befriend him. I felt like saying, "Bill, come
on, will ya? You got billions and billions. What do you say
you give me one billion? Do me a favor, okay? I'll owe you
one. What's it mean to you? I'll do things for you. I'll get
you girls, anything. I know the best massage parlors. You
won't be sorry. I know the right people. A friend of mine
runs an all-night crap game."

Nothing works out.
I bought an Apple computer.
There was a worm in it.

RIGHT: *Here I am alone with Bill Gates. The girls never*
showed up.

It's strange how getting up in front of a big crowd brings out fear in some people. To me it was never that difficult. I started in amateur shows when I was seventeen, and I've been doing it most of my life. But most people would fly to the moon before they'd stand up on a stage all by themselves and try to make people laugh. I know that for a fact. I was working at the Sands Hotel in Las Vegas, and after the show, Neil Armstrong—the first man to walk on the moon—came backstage to say hello. We shook hands and he said, "Wow, I wouldn't want your job."

I told him, "I wouldn't want *your* job either."

When I was a kid, I got no respect. I was kidnapped. They sent my parents a piece of my finger. My old man said he wanted more proof.

Back around 1995, people started saying to me, "Adam Sandler keeps talking about you in all his interviews." He would speak very favorably of me.

I'd never met Adam, so I was touched.

I finally got to know Adam when he asked me to be in his movie *Little Nicky*. I had a very small part, but Adam and I got to hang out a bit. He told me that when he was fourteen years old his father took him to see me at the Sunrise Theater in Fort Lauderdale.

Recently I was extremely honored when Adam presented me with the Comedy Idol Award at Comedy Central's award show, the Commies.

I like Adam, and I like his comedy. I think Adam and I are the only comedians today who have a clearly defined image. You can be a big star without having an image—Jim Carrey proved that—but Adam Sandler has an image that really works. In his movies he's a likable lug, and you're rooting for him when things go wrong. And when he finally wins at the end of the movie, he wins for everybody.

Adam gets involved in all the aspects of his movies, the writing, the directing, the casting, the producing, and all of his decisions are right on the money. I wouldn't be surprised if someday he owns his own studio.

Chapter Twelve
Stuck in a Bag
of Mixed Nuts

*Last time I saw a mouth like
that, it had a hook in it.*
—*CADDYSHACK, 1980*

When the credits roll at the end of my movie
Back to School, there's a line that reads
Estelle, thanks for so much.

I get a lot of letters wanting to know
who Estelle is and what I'm thanking her for, so instead of
answering every one of those letters individually, I'll tell
the story here.

When we opened Dangerfield's, my publicity man was
Richard O'Brien. Estelle Endler worked for him. After
four or five years, she moved with her husband to L.A. to
take a shot at doing publicity on her own. She was there
about six months when I got a call from her. She said,
"This agency, APA, would very much like to talk to you."

At that time in my life, I didn't want *any* hassles. I didn't
want to meet anybody. I didn't want to deal with money,
with problems. I didn't want any of the hassles that come
with life on the road. I didn't want to get caught up in the

business side of show business. I just wanted to stay in my club and work it like a nine-to-five job. So I told her I wasn't interested in meeting with APA.

"But they really want to talk to you," she said. "They think they can get you some great jobs."

At first I said, "I don't know." Then I said, "Okay."

So we all go into a room and talk. The APA agents tell me, "You're wasting your time just working at Dangerfield's. We can get you good jobs. We're talking big money—much more than you're making now in your club."

My kids were older now, so I finally said, "Okay, I'll give it a shot with these guys and see what happens."

They said, "Great. Who's your manager?"

I said, "I have no manager."

They said, "How can you not have a manager?"

I said, "I just don't. I just do things myself."

They said, "Well, we can't work with you unless you have a manager."

"Okay," I said. "I'll get a manager."

I turned to Estelle and said, "Wanna be my manager?"

She laughed. "I couldn't be your manager," she said. "I've never done anything like that."

"Are you kidding?" I said. "You're smarter than all of them. In two weeks you'll know more than they do. You're my manager, okay?"

She said, "Okay."

That's how Estelle became my manager. I didn't even ask her if she's a good driver.

I was right about Estelle—she was the best manager

anyone could ever have. She got me into movies.

In the late seventies, Estelle called me up one day and said, "There are these three young, funny guys doing a movie called *Caddyshack,* and they want to know if you'll meet with them."

So I meet the three guys who'd written the movie—Doug Kenney, one of the founders of *National Lampoon* and a cowriter of *Animal House;* Brian Doyle Murray, who played the caddymaster in *Caddyshack;* and Harold Ramis, the director. I liked them, and they liked me, so we made a deal. *Boom.* I get the job on *Caddyshack.*

It actually cost me money to do *Caddyshack.* I had to give up at least a month's work in Vegas. So it cost me $150,000 to do the movie, and they only paid me $35,000. People think I've made a fortune off reruns, merchandising, and stuff like that, but I got nothing: $35,000; that was it. My part in *Caddyshack* did get me into doing movies, though, so I guess it paid off in the end.

Early on, when we first started shooting *Caddyshack,* Estelle made a funny observation: "With the clothes you'll be wearing, it'll be like the whole movie is in black-and-white, and you're in color."

What a wild bunch that was—Kenney, Ramis, Brian and Bill Murray, Chevy Chase, Ted Knight. It was like being stuck in a bag of mixed nuts. None of us knew *Caddyshack* would be as big as it was, although whenever you start on a movie, you think you're making *Gone With the Wind.* Most of the time, though, the anticipation is far greater than the realization.

When *Caddyshack* came out, the reviewer in the *New York Times* said it was "immediately forgettable." Well, it grossed about $40 million at the time, and twenty years later, people are still repeating a lot of those "forgettable" lines. Here are a few of my favorite ones from *Caddyshack* written for my character, Al Czervik:

> *Hey, baby, you must have been something before electricity.*
> *He called me a baboon. He thinks I'm his wife.*
> *This is the worst-looking hat I ever saw. Looks good on you, though.*
> *When you bought that hat, did you get a free bowl of soup?*
> *Somebody step on a duck?*
> *Let's go while we're young.*
> *You wanna make fourteen dollars the hard way?*
> *My dinghy is bigger than your whole boat.*
> *Hey, everybody! We're all gonna get laid!*

Because of *Caddyshack* a lot of people think I'm an avid golfer, and I've probably been invited to play at every great golf course in the country. The truth is, I never play golf, and if I did, I'd probably be horrible at it, the worst. Not my kind of game. But I am sorry I had to give up those crazy clothes.

LEFT: Tell her I'll be waiting for her at the bar.

My golf game is getting real good.
Last week I got through the windmill.

---

*C*addyshack wasn't my first movie. The first movie I did was *The Projectionist* in 1969, the same year Dangerfield's opened. In the movie I played a theater manager who in the dream sequences became a bat.

Chuck McCann played my enemy. He was the good guy and I was the bad guy. In one scene Chuck was chasing me up a hill. I noticed he was having a tough time running. I had to slow down and look around bewildered to give him time to catch up to me. Finally we end up fighting on a cliff. After giving me a terrible beating for a couple of minutes, Chuck threw a dummy off the cliff. Not me, another dummy.

It was a low-budget movie.
We went to location by subway.

RIGHT: *When they said "punching bag," I thought I'd just be going a few rounds with my mother-in-law.*

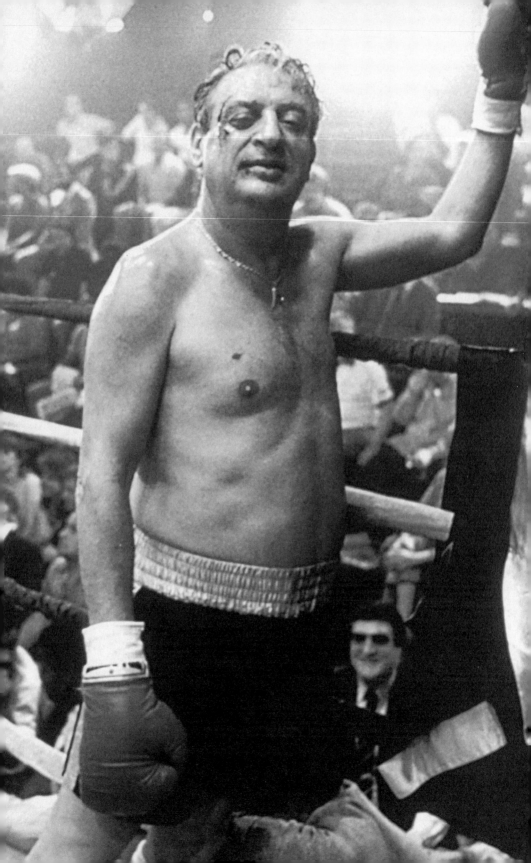

Thanks to *Caddyshack*, though—and Estelle, who passed away during the making of *Back to School*—I went on to make a lot of movies. Here are a few stories about some of them:

You were the inspiration for twin beds.
—EASY MONEY, *1983*

My next movie was *Easy Money*, which I cowrote.

One day, while we were filming on location in Staten Island, I had some time to kill between takes, so I was hanging with a couple of the limo drivers. After a few laughs and a few beverages, I had to go to the bathroom, but my trailer was all the way on the other side of the set, so the limo drivers suggested that I do what they did—use the bathroom in the funeral home across the street. They told me the man in charge was very accommodating.

So I walk across the street to the funeral home.

I walk in, say hello, and the man points me to the bathroom. I walk down a flight of stairs, make a left, and just before I get to the bathroom, I glance to my right. I was spooked when I saw a dead guy lying on a table.

LEFT: So you major in poetry? Good. Maybe you can help me straighten out my Longfellow.

When I'd finished in the bathroom, I went back upstairs, and—*boom!*—I see another body on a table in another room.

On my way out, I chatted with the man in charge for a few minutes. I told him I'd been a little surprised to see dead bodies there.

He said, "I'm having a good day. And I got one more coming in."

A woman named Thelma, who I mentioned earlier, worked for me for thirty years as a housekeeper. Actually she was much more than that. She helped me bring up my two children. She was family. She was a character and we had many laughs.

One day while I was shooting *Easy Money,* I got up early, about eight o'clock. Usually I don't get up until noon. Thelma says to me, "How come you're up so early?"

I told her, "I'm doing a movie."

She said, "Oh."

The next day, I got up early again. She said, "How come you're up early today?"

I said, "I'm doing a movie."

She said, "You did the movie yesterday."

One thing about my wife,
she gives great headache.
—BACK TO SCHOOL, *1986*

When you make movies there are always some hassles. When I made *Back to School,* there were some little problems and one really big one. When I made my deal with Orion Pictures, the writers who were going to write with me flew to Atlantic City, where I was performing.

We would write all afternoon and I would do my show at night. We were there for two weeks and then we went to Vegas. After two weeks in Vegas, we had the script written, so we went back to L.A.

The studio loved our script. I wasn't nuts about it, but I thought it was good enough. A little while later, I ran into Harold Ramis, who directed *Caddyshack.* Harold's a really nice guy and a very talented writer and director. I asked him to read the script and he agreed. The premise of the movie was that I'm a poor schmo who has to go back to college to motivate his kid. So far, so good.

So Harold read our script, and called me with a brilliant suggestion—he thought it would be a lot funnier if I went back to college as a rich guy. I agreed with him. Now I couldn't imagine doing the movie any other way, but I had to convince Orion to make a deal with Harold Ramis for a rewrite. They said, "What are you doing? We've got a great script here. Are you nuts?"

After two weeks, they finally came around and saw it our way. If they hadn't, the movie probably never would have been made.

Back to School took in nearly $100 million.

There is never a good time to get the gout. My first attack

hit while we were shooting *Back to School.* In the scene that day, I was supposed to talk to Ned Beatty, who played the head of the college (we called him "Dean Martin"), in his office. The gout really started to hurt my left ankle, so I lifted my leg to ease the pain and rested it on the chair next to me.

When Ned Beatty saw my foot there he said, in character, "Mr. Mellon, if you don't mind . . ."—insinuating that it was disrespectful for me to have my foot on his chair.

So I, also in character, said, "I'm sorry, Dean Martin," and took my foot off the chair and put it on his desk. I did that to keep my foot elevated, but it looked so funny we had to keep it in the movie.

I got lots of funny letters from kids after that movie. Here's one of my favorites:

Hi Rodney. My name is Tom O'Malley. I'm 10 years old. In the movie Back to School *you did a dive, the Triple Lindy. I say you really did that dive. My mother says you didn't. Who's right?*

I wrote back. I said, "Tom, listen to your mama."

RIGHT: Here I am with my kids, Brian and Melanie. My son, Brian, lives in Florida and loves to write songs. My daughter, Melanie, lives in New York with her husband and their newborn son, Joshua.

She does a lot of charity work.
She handles all the policemen's balls.
—LADYBUGS, *1992*

In 1992, I did a movie called *Ladybugs,* in which I coached a girls' soccer team. On the set, I met one of the executives of Mountain Valley Water, which I think is the best bottled water there is. We made a deal: I would publicize Mountain Valley Water in the movie. In return, they would give me Mountain Valley Water free for the rest of my life.

I like the Mountain Valley people very much, and I'm pretty sure they like me, but sometimes I wonder if they're happy that I'm still around. But then again, here I am, still plugging their water.

Now I'm gonna take your eye
out and show it to you.
—NATURAL BORN KILLERS, *1994*

LEFT: When Adam gave me a part in his movie Little Nicky, *I thought it was a reference to my anatomy.*

One afternoon, I got a call from the Academy Award–winning writer-director Oliver Stone, who wanted me to be in a movie he was about to shoot, *Natural Born Killers*. (At first I thought it was a movie about my wife's family.) Oliver explained the character he wanted me to play. It was a dark, twisted role—completely different from what I'd done in my previous movies—but I was interested. I asked him if the part was written yet. He told me no and he asked if I'd like to write it. I said okay and we made a deal.

If you saw the movie you know what a horrible person I played. In the film, I make it with my own daughter, played by Juliette Lewis, who is great. In one scene, I'm holding her and looking at her with a sick face. I say to her, "Go upstairs

Here I am as Monte, the baby photographer, in Easy Money.

and take a shower. Make sure it's a good shower, 'cause I'll be coming up later to see how clean you are."

The first time we did this scene, as soon as I said that line, Oliver yelled, "Grab her ass! Grab her ass!" I felt very strange doing that—she was a pretty girl and all, but I hardly knew her. Being a serious and dedicated actor, though, I grabbed her ass.

Just my luck, we did that scene in one take.

Sex after ten years of marriage . . .
Should the wife know about it?
—MEET WALLY SPARKS, *1996*

After *Natural Born Killers,* I went back to comedy. I cowrote *Meet Wally Sparks.*

Casting that movie almost got me evicted.

Since the early nineties, I've lived in Los Angeles. My new wife, Joan, and I live in a condo in a fancy building near Beverly Hills. Some of the residents are rather prudish.

One of the characters in *Meet Wally Sparks* was a hooker, so we had the actresses auditioning for that role wait in my lobby and then come up to my apartment and audition one by one. A lot of the girls came in character— that is, dressed as hookers. Sometimes two or three girls would be in my lobby.

Some of the tenants got very upset and complained to
the building manager and security, "There are prostitutes
sitting in the lobby!"

They thought these actresses were real prostitutes, and
that I was running them up to my apartment in shifts.

When my wife came home, the head of security and
the manager of the building pounced on her. They told her
about the tenants' complaints, and Joan explained that
the girls were actresses. That calmed down the manager
and the security guy, but I don't think all the tenants
believed her.

It's been several years since that happened, but some
of those people still look at me in the elevator like I should
be ashamed of myself.

---

*With my wife, I don't get no respect. The
other night there was a knock on the front door.
My wife told me to hide in the closet.*

Very often when Joan and I go out, she will drive the
car. After I am in the car, I realize I am uncomfortable.
My pants are tight on me, so I open them and pull down

RIGHT: Rover and me fixing our collars.

the zipper so that I have more room to relax while we drive around.

The problem is when we come back to our building. As we pull up to the entrance, the valet guys who take care of the cars see me pulling up my zipper, like something exciting just took place. They must wonder, *What's going on with these two?* In fact, I've had the feeling for quite some time now I'm the talk of the building.

It's our night! We're gonna paint the town yellow!
—ROVER DANGERFIELD, *1991*

I put some of my own money into an animated movie about dogs. It had some songs, which I wrote, and I even sang a few. In character, of course. I thought it was a funny movie, but I had some trouble with the studio, and they buried it like a bone.

In 2000, I did *My Five Wives,* which was based on an idea I got because my wife, Joan, is a Mormon. I started thinking that a movie about polygamy could be funny.

At the premiere of the movie, Joan and I renewed our wedding vows. Fabio was the best man and Adam Sandler was the ring bearer. It was a very classy party. The whole night only two fights broke out.

*I got lucky with one thing, I'm married
and I got a good woman. I got a woman who
loves me for my money and my fame and not for
what I am. She's a lovely girl. Her name is Joan.
She comes from Utah. She's a Mormon. I'm very
happy with her, very happy. In fact, next week
I'm marrying her sister.*

The *4th Tenor* came out in 2002, on my eighty-first birthday. In it, there's a scene in a restaurant where two people who are rather heavy sit down and start eating ferociously. I go over and say hello to them, they don't even look up. They just keep eating. No matter what I say, they won't stop.

Finally I say, "When you get to the white part, that's the plate."

We had to do three or four takes of that same scene. We did a close-up, a master, we had to repeat it, and through it all, these people just kept eating.

Then someone yelled, "Lunch!"

They were the first ones in line.

This pot should be good. I bought it from a cop.
—EASY MONEY, *1983*

Chapter **Thirteen**
I'm Not Going!

You don't know who to believe anymore. They say, "Love thy neighbor as thyself." What am I supposed to do? Jerk him off, too?

People always say to me, "Rodney, who makes you laugh?"

When I was a kid, Laurel and Hardy were my favorites, but I also loved the Marx Brothers, Jack Benny, W. C. Fields, and Mae West. I loved their images, and I loved their lines.

One of my favorite lines by Mae West: "Too much of a good thing can be wonderful."

When I was fifteen, I was in love with Henny Youngman. His act was laugh after laugh after laugh—*boom-boom-boom*. He'd tell a few jokes, play the violin, then tell a few more jokes. One of his best jokes was: "My wife and I, we're together fifty years. Where did I go wrong?"

Here's another: "I told the airline, fly me to Chicago and fly my luggage to Toronto."

They said, "We can't do that."

I said, "Why not? You did it before."

Youngman was also quick with an ad lib. I was in a nightclub in New York called the China Doll about sixty years ago, and Youngman was in the audience, listening to a girl sing. In the middle of her set, she said, "And now I'd like to take you on an imaginary trip to the Far East . . ."

Youngman stood up and yelled, "I'm not going!" And walked out.

I admire smart lines from anybody. This is from an Indian comedian named Charlie Hill: "They say Balboa discovered the Pacific Ocean. My people were living here for hundreds and hundreds of years. We never noticed it?

"One day the chief took his son to the top of a mountain. As they looked out over the hills and valleys, he spread his arms wide and said, 'Son, someday none of this will be yours.'"

Another comic I know, John Fox, a very funny guy, has a line he uses when someone heckles him. He says to them, "I'd call you a cocksucker, but I know you are trying to stop."

Bob Schimmel does a very funny line: "How does a blind person know when they're finished wiping their behind?"

Forty or fifty years ago, Birdland, a famous, wild jazz club in New York, had a black doorman named Pee Wee. He was a very short guy, but he had a way about him. He was a bit snippy, and he walked around like he owned the place.

I saw him get into an argument one night with a customer. They exchanged some heated words, and the guy said to Pee Wee, "Don't you bug my ass, you half-a-motherfucker."

We all like different types of shows on television. I like a show where anything can happen. *The Howard Stern Show* does it for me. I wish I was Howard Stern. He has a way with women. Whatever he tells them to do, they do. He says, "Pick up your dress, honey. Higher, higher. I want to see your ass. Higher." They do it. Girls don't listen to me that way. I go out with a girl, spend all kinds of money. She won't take off her gloves.

I tell ya, my wife and I don't think alike. We got problems. I want to go see a marriage counselor and she wants to go on The Jerry Springer Show.

A show that really makes me laugh is *The Jerry Springer Show*. It's much funnier than all the sitcoms with their piped-in laughter. The show is real, the people are real. There's nothing better than something funny that's spontaneous, something that comes out of the moment, instead of out of a script.

I want to go on The Jerry Springer Show,
but they turned me down. I got all my teeth.

For some people, the thrill of gambling is better than sex. If you don't believe me, just hang around the slot machines in Vegas for a while. A woman wins five dollars and she screams like she's having an orgasm. If a woman was being attacked in a casino with slot machines and she was yelling and yelling, the guard wouldn't do a thing.

He'd just think, *Another winner.*

People go crazy to make money. I guess they wanna see how much they can die with. But like Redd Foxx said, "I never saw a Brinks truck following a corpse."

People think that the more money you have, the happier you'll be. Then why does Connecticut, the richest state in the country, have the highest suicide rate? So if you want to live a long time, stay broke.

*I tell ya, in Vegas you gotta go broke. They got
slot machines all over. Even in supermarkets.
I bought a container of milk—cost me $238.*

*RIGHT: Here I am in Vegas making a few extra bucks
holding up a couple of billboards.*

Back when I was selling aluminum siding, I was in a customer's house one day, chatting up the lady of the house, who was standing there with her dog and her two small children. I said to the woman, "Cute dog."

"Yeah," she said, "he's very cute." Then she pointed to her children. "If it wasn't for them, I could spend more time with him."

Children have their own way of looking at things. I was walking on the beach one day when a group of kids came running toward me, asking for my autograph. So I signed something for each of them, and they ran off.

The last kid was a little girl, ten or eleven years old. As I was signing my autograph for her, she said, "You're doing pretty good today, huh, Rodney?"

I said, "Yeah, I guess so."

She thought I was out there trying to sign as many autographs as I could.

I shouldn't tell jokes about my wife.
She's attached to a machine that keeps
her alive . . . the refrigerator.

I've learned a lot of things. One is never have dinner at a friend's house. From the husband you hear things like,

"My wife's the best cook in the world." My last dinner at someone's house did it.

My friend and I sat at the table while his wife was serving the food. There were some chicken wings on a small plate. I started nibbling on one of them while waiting for the main dish. Then I said to my friend, "What's the main dish?" He said, "Chicken wings."

I was in shock. I said, "Chicken wings?"

He said, "Is something wrong?"

"No, I love chicken wings," I said. "Little crushed bones with pounds of fat rolled around them. Why, chicken wings, that's my favorite."

My wife can't cook at all. In my backyard,
the flies chipped in to fix the screen door.

You always hear that you get wiser as you get older, but the longer I live, the less I understand.

In my travels and with all the years I've spent hanging out in clubs and bars, I've spoken to many married men. I've learned that when they do something that is considered wrong, they justify their actions with some twisted reasoning. And I've heard all the reasons they give to justify cheating.

I'm sure you've heard the standard ones: "As long as

my wife and kids are provided for, then I can do whatever I want."

One guy was good at mathematics. He told me, "Whatever I earn, two thirds goes to the family and one third goes to me."

Or, "I cheat to see if *all* women laugh during sex."

And, "I cheat so that I can get a decent breakfast."

I've had married men tell me, "I never come on to a girl. *That's* cheating. But if a girl comes on to me, that's not cheating. And if I knew a girl before I knew my wife, then that's not cheating."

One guy told me he never cheats on his wife—or his mistress.

And we've all heard this classic justification for cheating. A guy's wife could be a wonderful person, a churchgoer who helps the poor, and he makes a tramp out of her with the implication: "Who knows what *she's* doing?"

I tell ya one thing, my wife keeps me in line.
No matter how many guys are ahead of me.

The best justification for cheating I ever heard was from a guy I met at a hotel in Las Vegas. We were sitting at the pool, and his wife and two kids were splashing around in the water.

I said, "Your wife is a lovely woman."

He said, "Yeah, she's good. I love her. My kids, I love them, too. I tell ya, I've got a beautiful family."

I said, "Do you play around?"

He said, "Oh yeah." He told me that when he plays around, he does it for his wife. "If I didn't play around," he said, "I'd be miserable to live with."

I went out with a hooker. She told me, "Not on the first date." So I saw her again. This time, she drove a hard bargain. She said, "The sex will be seventy-five dollars." I said, "I only have fifty dollars." She said, "Okay, I'll do it for fifty. But I'm telling ya, I'm not making a dime on you."

I put on TV the other night and I came to a fashion show. It puzzled me. All those beautiful models walking around and they all look mean. Why don't they smile as they walk? Who are they mad at? They do all right. They make good money. All the guys love them. All the girls want to look like them. But they still walk around mean. It's a mean turn. A mean stop. It's always a mean face. They make me feel like I did something wrong.

Another thing that puzzles me. When models pose for pictures, they show their belly button. Why? A belly button

is not sexy. A belly button is good for only one thing: when you're lying in bed eating celery, it's a place to put the salt.

Here's something I don't understand. When I got married, the guy said to me, "You may kiss the bride." Big deal. After all the things I've already done to the bride, he tells me I can kiss her.

When I watch a football game, I see guys trying to bang the other guys as hard as they can. They tackle hard. Their heads collide. Their bodies slam against one another. And all of a sudden the game stops. There's a penalty for "holding."

And another thing.

Why do they make such a big deal out of the "two-minute warning"? Everyone knows you got two minutes to play and that's it. To me, a two-minute warning is like when I'm in bed with a chick. The phone rings. It is her husband calling from his car phone. He says, "Honey, I'll be home in two minutes." Now, *that's* a two-minute warning.

I also don't like when they have girl announcers for a football game. They should have only male announcers. Football is a man's game. I don't want to hear a girl tell me it's two inches short.

LEFT: *When I say she's a doll, I mean it.*

*With my wife nothing comes easy. When I
want sex she leaves the room to give me privacy.*

I was working in Atlantic City. One night after the show, my friend and I went to a little nightclub to get a bite to eat. We were feeling good, had a few drinks. There were two girls who worked onstage there. One played the piano, one played the harp.

Not too many people were there. It was toward the end of the evening, and the girls were close by. So we started talking to them. It got to a point where they were coming off in about fifteen minutes, and maybe we'll go somewhere and have a drink.

They said, "Fine." So we had a date.

My friend says to me, "Which one do you want?"

I didn't know which one to pick. They were both attractive—maybe the harp girl had a slight edge.

Then I thought, *She plays the harp. It seems like such a religious thing, a saintly thing, a "do what's right in life" thing. The chances are she's not gonna be a wild girl.*

I picked the piano player.

As usual, I picked wrong.

*I know how to always make a woman say yes.
I ask her, "Am I bothering you?"*

Chapter **Fourteen**
Three Lucky Breaks

I'm at the age now, when I
meet a woman sixty years
old, she's too young for me.

L et's face it, I'm getting old. That's bad enough, but in the last few years, I've had four major operations. I've been cut up so many times, I feel like I'm back in my old neighborhood.

Before each operation I've had, the same thing always happens. As I'm lying on the gurney, the doctor comes over and he smiles at me. The smile says, "Don't worry. I know what I'm doing. Have confidence in me."

I always tell the doctor, "If I don't make it, I'll never know it."

My first major operation was in 1992, to fix an abdominal aortic aneurysm. That was serious—if your aorta goes, you go. Most people die in minutes. Lucille Ball, George C. Scott, and Albert Einstein all died of ruptured aneurysms.

When a doctor catches it in time, it's often discovered by accident. That's what happened with me. One day I woke up and had some pain on my right side. I went to the doctor and got an X-ray. As the doctor had suspected, the tests showed that I had pancreatitis. But they also showed something the doctors really didn't like—an aneurysm. So I had the surgery—what they call "the open procedure."

Here's how that goes: First they cut you open from your diaphragm down to your "ecstasy rod." Then they take all of your intestines and put them on a table next to you. Then they perform the operation.

When I came to, I was in intensive care. My torso was wrapped in bandages, and there was an IV stuck in my arm to feed me intravenously.

A doctor looked in on me and said, "Hiya, Rodney, how ya doin'? Don't worry, we'll have you walking in no time."

He was right. I got the bill. I had to sell my car.

I mean, I'm not a kid anymore.
I could go tomorrow. And I hope I go
tomorrow. I haven't gone today yet.

People often say, "It's a miracle I'm alive." And for me, they may be right. I was a heavy smoker for over fifty years. Never could stop. I used to walk around with three different packs of cigarettes in my pockets—filters, nonfilters, and menthols. Sometimes I'd quit for a whole day. Then I'd give myself a reward—a cigarette.

That's how I ended up going to the Pritikin Longevity Center in Santa Monica in 1982. It's no longer there, but it was a highly recommended place where you could lose weight and stop smoking. So I thought, *I'll check in there for a month. I'll take care of myself*—which I did.

One morning, I ran into the head man, Nathan Pritikin, a great guy who was really down-to-earth.

I said, "Dr. Pritikin, nice to see you."

He said, "Rodney, how's your cholesterol?"

I thought, *Wow, he gets right to it.*

"Getting better, Doc," I said. "But tell me something. You say don't eat lobster because it's all cholesterol. But if lobsters are all cholesterol, how come they live a hundred years?"

Nathan had a sense of humor. He said, "They don't smoke, they don't drink, and they watch what they eat."

I said, "How about sex?"

He said, "No thanks. I don't know you well enough."

After a month in that place, I felt like a tiger.

After I left Pritikin, I didn't smoke for three years. It gave me a chance to clean out my chest. It made my lungs fresh. Then one night I got drunk, and I had a cigarette.

After that, I was back smoking again. I smoked even after I had my aneurysm operation, right up until my double bypass. By then I knew it meant my life. So I finally stopped.

I haven't smoked now in over three years. After my last

bypass operation, I was all cleaned up, and I'd be a fool to start stuffing my lungs up all over again with cigarette smoke.

I never thought I could quit, but I did. Now when the urges come, when I think I can't make it, I just remind myself that nobody ever died from not smoking.

What a doctor I've got—he's really mixed up.
Last week, he grabbed my knee and told me to cough.
Then he hit me in the balls with a hammer.

I got three lucky breaks that summer I spent at Pritikin. I lost weight, I stopped smoking, and I met my wife, Joan Child. She owned a flower shop in Santa Monica. One day I stopped by to smell the roses. And I stopped by again the next day. And the next day. And the next . . .

We started dating, which was tough because I wasn't in L.A. much once I got out of Pritikin. At that time I was doing dates all over the country and in Canada. Luckily, we stayed at it, though; we dated for ten years.

One night, when Joan was closing up her store, she said, "I know a great place to get a bite."

RIGHT: *I went back to visit all my old schoolteachers.*
I only had to make one stop, the cemetery.

Smoking Section

-- Rodney Dangerfield

I said, "Let's go. But I need to take my car, too. I have to get up early tomorrow, so from the restaurant, I'll go right back to Pritikin."

"You'll never find this restaurant," she said. "Follow me."

We took off in two cars. I was right behind her, but I had to drive fast to keep up. Next thing I knew, there was a police car behind me, lights flashing, bullhorn screeching, "Pull over!"

But I couldn't. I knew that if I pulled over I'd lose her, and I didn't know where we were going, so I hit the gas—and so did the cops. So now I was following Joan, and the cops were following me, lights still flashing.

When Joan finally stopped, the cops pulled up with their sirens wailing, bullhorns, the whole thing. I thought they were going to drag me out of my car and club me to death, but they were cool. When I explained the situation, they started to kid around with us, they were all right. Then they said they wanted my autograph—on a ticket.

People think I get plenty of girls.
I go to drive-in movies and do push-ups
in the backseat of my car.

LEFT: *Me and my beautiful wife, Joan, on our wedding day.*
Joan is a Mormon, so next week, I'm marrying her sister, too.

I moved to L.A. in 1990, and Joan and I finally got married on December 26, 1993. It wasn't a big, dramatic thing— we just decided to do it. We flew up to Vegas, got married, then I played some craps and we flew home that night.

Besides going into show business and opening Dangerfield's, this was another time people told me I was nuts. I married Joan with no prenuptial agreement. They thought I was making a big mistake. We are now married over ten years. It looks like "the mistake" worked out.

I learned in life, you can never say never. After my first marriage, I said I'd never get married again. Here I am married again, and this time it is the way it should be. Joan and I get along great. Thanks to Joan, I am in love, and I'm loving it.

I still have my bouts with depression, but I work through them. The worst depression I had was when I was in my seventies. It was a bad one. For two years, I couldn't function. Finally I snapped out of it and started working again.

When we got married, the first thing my wife did was put everything under both names—hers and her mother's.

LEFT: *My neighbors complained when I tied up traffic to have my new hot tub flown in. I didn't see what the big deal was—it's not like I was naked in it.*

In 2000, I had a double bypass, which is a technical term for: they cut my chest in half with an electric saw. It still hurts when I think about it.

When the operation was over and they wheeled me out, Joan was waiting for me in the cardiac intensive care unit. She later told me I had tubes stuck in me in every imaginable place, which reminds me of something my friend Joe Ancis said: "When you come to the end, you have a pipe in your nose, a pipe in your mouth, a pipe in your chest, arm, and neck. For the finish, we all turn Scottish."

Later that night, when they took the tube out of my mouth, I could finally talk. I looked at my wife, and all I could say was, "Pain."

Joan went to the nurse and said, "Look, he's in a *lot* of pain. What can you do for him?" So they gave me something. Within five seconds the pain was gone and I was smiling. It was fantastic.

I asked Joan to find out what it was that they had given me. She told me it was Dilaudid—synthetic heroin—and boy, I can see how people get hooked on the real stuff.

I tell you, with my doctor, I don't get no respect.
I told him I'd swallowed a bottle of sleeping pills.
He told me to have a few drinks and get some rest.

I was lucky I made it through all my surgeries. Many aren't so lucky. Hospitals now admit that 120,000 patients a year die because of their mistakes. (It's probably much more than that, but that's all they will admit to.) That comes to more than 300 people a day who die by mistake. When you read things in the paper like *He died after complications from heart surgery,* that's a lot of BS. They probably accidentally killed him.

You see a doctor. He looks so mature. His hair is neatly combed. His pants are perfectly pressed. In his pocket he has two pens. He's smart. He knows that if one don't work, he's got another one. It's hard to imagine that an hour from now there's a chance he'll be killing someone.

I've experienced plenty of hospital mistakes. Most of the nurses who brought me my pills made errors. Many times Joan stayed with me overnight in the hospital just to check on the pills—and they were usually wrong.

My doctor was giving me a complete physical, so he said, "I want a urine sample, a stool sample, and a semen sample." So I left my underwear and I went home.

The doctors aren't the only ones who don't always prescribe the proper medication. I can do that for myself. I was in St. John's Hospital recently. I forget exactly what was wrong, but I wasn't feeling too good, so I went down there to get an EKG and a checkup, and it was a drag. I ended up in intensive care.

I was really bored, waiting around for hours, so I thought, *Hey, there aren't too many people here, and it's dark. I'll light up a joint. Nobody'll notice, and I'll feel okay.*

The smoke from the joint went right down the hallway—a breeze or the air-conditioning must have caught it or something—and everybody could smell it. Two minutes later, a security guard came over.

I got lucky, though. He was a nice guy. I told him, "My wife won't let me smoke at home, so I decided to come over here."

You know you're getting old when your insurance company sends you a half a calendar.

Chapter **Fifteen**

Turkeys in Wheelchairs

*I got no sex life. My dog keeps
watching me in the bedroom.
He wants to learn how to beg.
He taught my wife how to
roll over and play dead.*

've met a lot of funny people in my life, but to me and
for those who knew him, the funniest guy going was my
friend Joe Ancis. A lot of people say Lenny Bruce was
influenced by Joe's comedic genius.

You've probably never heard of Joe because he wasn't
famous. He had no desire to get up on a stage or be in
show business in any way.

I knew Joe for over fifty years. In fact, he lived with me
for eighteen of them. He died while I was writing this
book, and I can't believe he's gone, because we always
talked about how I would be the one to go first. I was the
wild man, the drinker, the smoker. Joe was extremely careful, probably too careful.

Most of Joe's best stuff was of the "you had to be

there" variety, but I'll try to lay some of his things on you.

When Joe lived with me, I had a miniature poodle named Keno. One time I was talking to Joe, but I was distracted because I noticed that the dog kept looking at me.

After a few minutes, I said to Joe, "What's with the dog? He keeps staring at me."

Joe said, "Man, you're a star."

One day Joe said, "How's the weather outside?"

"I don't know," I said. "Go out on the terrace."

"No, they'll want speeches," he said.

Back in the fifties, Joe and I worked together in the siding business in Englewood, New Jersey. He was a salesman— a very good one—but like me, he never got up early, which was actually okay because one of the biggest advantages of both the siding business and show business is that you can sleep late. So we never made appointments in the morning—we always got up late and went to work late. Sometimes, though, a client or a prospective customer would try to schedule an appointment in the morning.

I heard Joe deal with that situation on the phone once. Joe said to his customer, "Tomorrow morning? Uh, let me see . . ." He pretended to check his date book. "Well, tomorrow morning I'm all tied up," Joe said, then put his hand over the phone, looked at me, and said, "In my pajamas."

Another time Joe and I were watching a boxing match on TV. Joe said, "If only one of them would just say, 'I'm sorry.'"

One of my favorite things to eat is a turkey leg, and I have that all the time, at home or in restaurants. One night Joe saw me eating a turkey leg. "Man," he said, "you've put more turkeys in wheelchairs . . ."

Joe was paranoid about germs, so he had his own way of using public restrooms. He would get into a stall, lock the door, then stand on top of the toilet seat, take all his clothes off—except for his shoes—and squat over the toilet. (He told me he had to be very careful when doing this because he didn't want any splashes.)

One time Joe went into the stall but forgot to lock the door. He was right in the middle of his routine—standing on the toilet seat, naked, facing the door—when some guy looks under the door, sees no feet, and figures the stall is empty.

This guy opens the door, sees Joe—now seven and a half feet tall, and naked—screams, and runs out of the men's room.

Germs weren't the only thing Joe was afraid of. His parents made him fear many things. When Joe was a kid, his father told him, "Never fly on an airplane. They explode in the air." That was just one of many nutty things Joe was told growing up. Joe wasn't much for dating because his mother once told him, "If you break up with a girl, be very careful. They throw acid in your face."

Joe would never sleep with the window open because he was afraid a pigeon would get in and peck him in the eye.

The following could only happen to Joe with his head full of fear. It happened forty-five years ago.

Joe was very much into music and singers. So one

night he went to see Nancy Wilson in concert in New York. Halfway through her act, he noticed some empty seats way up front. During the applause for Nancy's next song, he walked to the front, around the third row, and took one of the empty seats.

Then his head went to work.

Joe started thinking, *What if someone shoots Nancy Wilson? They'll think I did it. I'm the only white guy in the place. They'll say, "It's him! He even changed his seat to be close and get a better shot. He wasn't even applauding for Nancy."*

For the rest of the performance, Joe sweated it out, hoping that Nancy didn't get shot. Joe kept applauding heavily for Nancy and making sure people sitting near him would see him smiling. That was Joe's head. It was tough for him to relax.

One day I was in a bank with Joe when he wanted to cash a check. The woman behind the cage looked at Joe and said, "How do I know you're Joe Ancis?"

Joe said to her, "How do I know you're Next Window Please?"

I went to buy a suit. I told the salesman "I wanna see something cheap" . . . he told me to look in the mirror.

RIGHT: Joe Ancis, my best friend, was the funniest guy I ever knew.

Joe and I understood each other, and we had the same kind of dark thoughts, so we got along great. But unlike me, Joe never wanted to meet people. He used to say, "I'm not lookin' to make new friends—I'd like to lose the ones I got." Me, I love communicating with people. I'd rather be in Secaucus with somebody I can talk to than be in Paris alone.

People always wanted to meet Joe, though. He was famous among the hipsters and comedians, who'd heard so much about him, but he didn't want to meet them.

Dave Goldes, who wrote for Carson and for me, was always bugging me, saying he wanted to meet Joe. I'd say, "What can I tell ya? He's not lookin' to meet you."

One night Joe and I were eating at the Stage Deli when Dave walked in. Dave spotted us and immediately headed our way, thinking this was his big chance to meet Joe. I saw Dave coming, so I said to Joe, "Look, man, here comes Dave Goldes. You've heard me talk about him. He's a great guy. How 'bout letting him sit with us for a minute, okay?"

Joe said nothing.

When Dave got to our table, I said, "Joe, say hello to Dave."

Joe's first words to Dave were: "Your mother sucks midgets."

Dave had heard so much about Joe that he broke up laughing. I thought it was funny, too, but the midgets at the next table dropped their corned-beef sandwiches in shock.

A few weeks ago I did a show. The whole audience was midgets. I got a standing ovation and I didn't even know it.

One day Joe and I went to the beach. As we were getting our stuff out of the car, a cop said, "Hey, Rodney, how ya doing? Did you come for the nude beach?"

"Nude beach?" I said. "Where?"

The cop said, "Right over there."

So Joe and I took a walk on the nude beach. We didn't have to get naked, but everyone else there was, which was fine by us.

We were strolling along the water's edge, pretending to mind our own business, when we saw this man walking toward us, like he was in a hurry. He looked to be about sixty years old, gray hair, very mature, with his dick swinging back and forth.

Joe looked at me and said, "Monday morning in the bank, that same guy turns you down for a loan."

*Last week, I had a bad experience.
I went to a nude beach. They kicked me out.
They told me it's impolite to point.*

Forty years ago, I was feeling really depressed, even more than I usually do, so Joe recommended a famous psychologist to me. I went to see this guy many times, and he was very helpful. I still remember two things he told me: "People are all fucking crazy and most are unethical." Like I said, a smart guy.

Joe had a similar take, but he put it brilliantly. He said, "The only normal people are the ones you don't know too well."

I know I'm getting old. When I whack off I get tired holding up the magazine.

Chapter **Sixteen**
My Heart Started
Doing Somersaults

*I told my doctor, "I think
my wife has VD." He gave
himself a shot of penicillin.*

Although I complain about doctors, I'm only alive today because of some very good ones. I was having some unusual symptoms around the time of my eighty-first birthday in November of 2002. I was only able to walk short distances before becoming exhausted, and Joan thought I might have been having little strokes—I'd sometimes say something odd or disconnected, totally left-field kinds of things. How she could tell the difference, I'll never know.

We knew I had a bad aortic valve that was considered too risky to repair when I had my double bypass in March 2000, and we attributed most of my symptoms to that.

In addition, my right carotid artery had been completely closed up since New Year's Day of 1991 when I suffered a tiny stroke in Florida. But by February 2003, my shortness of breath was worsening and my doctors were debating whether or not to go ahead with the aortic valve replacement, despite the risks. The way they explained it, I might suffer a stroke during the surgery or go into heart failure if I didn't have the surgery soon.

A short time later, my heart started doing somersaults in my chest, and I had to spend a night in St. John's Hospital in Santa Monica until they got it under control. My cardiologist told us that I would only live three months if I didn't get my valve replaced.

My doctor is strange. No matter where it hurts you, he wants to kiss it and make it better. After he checked me for a hernia, I had to change my phone number.

Joan and I quickly started interviewing heart surgeons and decided to go with Dr. Hillel Laks at UCLA Medical Center. When I went to him, he ordered a bunch of tests on me, including angiograms to my brain, which revealed a whole boatload of problems. Not only was my right carotid 100 percent blocked, but so was my right vertebral artery. Once those suckers close off, you can't do much about it. The aortic blockage was moderate to severe and one of the bypass grafts from the previous heart surgery had closed down. The doctors said that either the aortic valve or the blocked neck arteries or a combination of both could be causing my symptoms, and they needed a little time to evaluate the situation.

Here's a free anatomy lesson, and not the kind I had to pay for in those Times Square strip clubs: four arteries supply blood to the brain. I was down to two and they were both on the left side of my neck. That meant the right hemisphere of my brain must have been very blood-thirsty! The two arteries on the left were no day at the beach either—they had partial blockages in six places. The doctors said they couldn't stent them, so heart surgery would be very risky because there was a good chance I'd have a stroke in the middle of it.

Joan and I met with Dr. Laks a couple of times. He suggested that I get examined by Dr. Neil Martin, chief neurosurgeon at UCLA, before going ahead with the heart surgery.

Dr. Martin ran me through another marathon of tests and, after looking at the results, said the words "brain surgery" to us for the first time. He said the good news was that he happened to be a specialist in the very procedure that could protect me from having a stroke during heart surgery. He called it an extracranial, intracranial bypass. I only know this because Joan had him write it down. When we got home, she immediately got on the Internet to study up on it.

Meanwhile, Dr. Martin told us that he wasn't sure he would recommend the surgery because he wasn't certain my heart was strong enough to endure it. He wanted to talk some more with Dr. Laks, and said he would give us a call in a day or two. So Joan and I went home not knowing if I was going to have brain surgery or heart surgery, or both. Some choices, right?

My doctor told me he had good news and bad news. I said, "Doc, only tell me the good news, all right?" He said, "All right. They're going to name a disease after you."

The following day, which was a Saturday, our building concierge took a message from the brain surgeon that read: *Spoke to Dr. Laks. Decided against heart surgery. Go ahead with brain surgery.*

This really surprised us, so we quizzed the guy who took the message to be sure he got it right. He was sure, though, so we prepared ourselves for that.

We weren't able to reach the doctors until Monday, when they told us that the concierge had taken the message down backward. I would be having *only* the heart surgery, which would be scheduled as soon as possible.

By this time, I was not able to stand for more than about ten seconds without experiencing symptoms. I could barely get from one room to another, so Joan rented me one of those three-wheel scooters so I could get around in our apartment.

PHONE CALL

FOR _MRS. DAngerfield_   DATE _3/8_   TIME _12_   SAT. 30 A.M. P.M.

M _DR. Martin_

OF

PHONE/MOBILE _(310) 428-8232_ FAX

MESSAGE _Spoke to DR. Laks._
DECIDED AGAINST
HEART SURGERY.
GO AHEAD WITH
BRAIN SURGERY
SIGNED _CAll Him._

1154

☑ TELEPHONED
☐ RETURNED YOUR CALL
☑ PLEASE CALL
☐ WILL CALL AGAIN
☐ CAME TO SEE YOU
☐ WANTS TO SEE YOU

To show you how serious all this was, I even stopped smoking pot to clear my lungs. Joan had me eating healthy foods and getting lots of rest. It was much worse than when I checked into the Pritikin Center. No sneaking off for Chinese food with Roseanne.

A few days later, the date for my heart surgery was set. I had to be admitted the following afternoon, so my son and daughter flew in from the East Coast. I was scheduled for another angiogram to my heart to make sure there were no surprises before surgery the following morning.

That day, a lot of my friends called to wish me luck. I

ABOVE: Here's a message I hope you never get.

can't remember much of that time, but Joan says Jay Leno, Jim Carrey, Adam Sandler, Louie Anderson, Dom Irerra, Brad Garrett, Lenny Clark, Michael Bolton, Robert Davi, and Bob Saget were some of the more famous ones.

That night a big group of us were having dinner in my hospital room, when Dr. Laks walked in with a few other doctors and asked to speak to Joan and me alone. When everybody had cleared out, Dr. Laks looked me in the eye, man-to-man like, and said, "Rodney, you like to be creative, don't you? Your work requires a lot of thinking, so we want to do everything we can to preserve your brain."

Uh-oh. I started hearing that damn *Twilight Zone* music in my head. *Do-dododo, Do-dododo, Do-dodo . . .*

Dr. Laks continued, "Our team has reevaluated your case. Your angiogram this morning gave us some new information. We feel your heart is stronger than the initial tests revealed."

Then I realized what he was saying: they wanted to do the brain surgery *first*, then do the heart surgery. They had concluded that my heart would be strong enough to endure the brain surgery after all, and that would optimize my chances with the heart surgery.

We figured the brain surgery would take place the following day, but Dr. Martin was on vacation and wasn't reachable by phone. (I think he went mountain climbing or something.) I didn't want to spend Dr. Martin's vacation in

LEFT: I got an advanced degree in making advances.

the hospital, so Joan took me home and I continued to practice parallel parking the scooter in our dining room. I was still writing jokes and working on this book.

I told my doctor, "Every day I wake up,
I look in the mirror, I want to throw up.
What's wrong with me?" He said, "I don't
know, but your eyesight is perfect."

The big day finally came. On April 7, 2003, I checked into the hospital, this time for brain surgery. My son and daughter flew in again. That night we had a small party in my room. Bob Saget and Louie Anderson kept us all laughing until we were exhausted.

Joan spent the night with me and went to pre-op with me in the morning. Just before they gave me the injection to knock me out, Joan says I said, "I want to live . . ."

Those were my last words for about two weeks.

My doctor's a very strange man.
I said to him, "Doc, what's the difference
between an oral thermometer and a rectal
thermometer?" He told me, "The taste."

The surgery went well, but there were still some spooky days ahead. For various reasons, they had to place me in a "medically induced coma," which, they said, gave me a better chance at a full recovery. Since I was in a coma for part of this time, and pretty whacked out for the rest of it, I don't have many memories of what went on, so I'll turn things over to Joan for a while:

When Rodney was taken out of his coma, he was still pretty far gone. He couldn't talk, or hardly move, so I was always looking for clues to figure out if his unusual and newly blood-soaked brain was processing information. Some days he would seem responsive. Others not very. One day his eyes were open and he was able to weakly squeeze my hand, but not really on command.

Dying to know if he could understand what I was saying to him, I finally asked him a question I figured would get a reaction. A while back a fan had e-mailed Rodney an X-rated cartoon that Rodney thought was hilarious. I can't bring myself to describe the cartoon, but the caption was: Surprise Balloon. So, that day, I asked Rodney if he wanted to see a surprise balloon. What a reaction! He grinned from ear to ear and tried to talk. I'll never get that picture out of my mind—both sides of his mouth turned up in a smile despite the big ventilator tube stuck down his throat, with another one up his left nostril, and his skull shaved and covered with staples and electrodes. I was thrilled.

A few days later, another thing happened that assured me Rodney was on the road to recovery. Before his surgery we always looked forward to one daily ritual, watching The Jerry Springer Show. *This went on for years and years. We never missed that show, no matter what was going on.*

The morning they finally took all the tubes out of Rodney, he was sitting up in bed in the intensive care unit, and I noticed that he was staring at the clock and then looking out the window. It took me a minute to realize that it was 10 A.M., which meant that Jerry Springer *was on. I asked him if he'd like to watch, and he said* yes *in this deep, almost satanic voice, so I turned it on very quietly, because I didn't want to disturb anybody else in the unit, or let the doctors know what we were doing. When Rodney saw* Jerry Springer, *he lit up. He was so into it—like a kid making his first trip to Disneyland. Pure glee.*

The word quickly spread that not only was Rodney okay, but that the first thing he had wanted to do when he had regained consciousness was watch Jerry Springer. *Our publicist mentioned that in the press release that day and the story was picked up all over. Jerry Springer even sent us tapes of all the shows Rodney had missed while in his coma.*

Oddly enough, though, once we were home, the fully recovered Rodney no longer had any desire to watch television. But for a while, especially with restricted blood flow to the brain or while coming out of a coma, Jerry Springer *was a wonder drug, a tonic for Rodney's tortured soul.*

It turns out the UCLA doctors are geniuses. The brain surgery was the right way to go. My recent echocardiogram shows normal heart function. In other words, I got to skip having heart surgery for a while. I still have aortic valve stenosis, but I don't have shortness of breath. A full year after the surgery, I'm breathing fine, walking fine, thinking sharp. One of these days I will have to get that aortic valve replaced, but for now, I don't think about it.

By the way, originally I was going to include the Surprise Balloon cartoon in this book but the publishers felt it was too risqué. If you want to see this X-rated cartoon, come to my website, *www.rodney.com,* or send me an email to *rodney@rodney.com.*

I tell ya, I think doctors get too much respect.
A hooker should get more respect. She's
more important than a doctor. I guarantee you,
at four o'clock in the morning, drunk, I'd never
walk up five flights of stairs to see a doctor.

Chapter **Seventeen**
End of the Line

With me nothing comes easy.
This morning I did my push-ups
in the nude . . . I didn't see
the mousetrap.

People ask me, "How do you write jokes?" There is no set procedure, but writing is basically thinking. Before you write anything, you have to sit there and think about it. Sit down and try to think funny in whatever area you want: wife trouble, car trouble, kid trouble.

With my image, *everything* is trouble.

Many of my jokes are written on the spot when I hear something I can make funny. One day I was sitting in Mulberry Street Pizza in Beverly Hills. It's a nice place to hang out with the guys—and one of the guys I was hanging out with at the time said that he had a "tale of woe." So I turned that into: "Every man has his tale of woe. Unfortunately, in life there's more woe than tail."

One time I was kidding around with a waitress at

Mulberry Street. I said, "Make it with me, and I'll give you a thousand dollars for an hour."

"I'm sorry," she said, "that's not enough."

I said, "Okay, I'll stay two hours."

I was in Fort Lauderdale, driving along by myself. I came to a red light, and at the intersection was a convertible with two young couples. To my left, about three feet away from me, was a very attractive girl. She looked at me in a sexy way and said, "I wanna suck your cock."

I could see what her game was—she wanted to see if she could shock the "old man."

So I said, "You're gonna have to pay me."

As I drove away, I could hear them laughing and yelling, "Rodney!"

Putting a good joke together is a delicate thing. The emphasis on the right word is very important. So is the rhythm, the timing.

And of course, the joke has to be funny.

People don't know the preparation you do in show business. I still do *The Tonight Show with Jay Leno,* and I try to do all new jokes each time.

Counting my stand-up routine and my "conversation" on the couch with Jay, I need about 30 new jokes. That

LEFT: *Here I am with Jim Carrey, Bob Saget, and some of my other favorite young comics. Who the fuck knows where we were?*

means I have to write a hundred new jokes. Then, to know which jokes get the best laughs, I go to a local comedy club the Laugh Factory and try them out.

After I have about 30 new jokes I like, I have to put them in some kind of order—create some flow, some continuity. I call that stringing them together into my joke necklace—they have to be in the right order to work. I have to string together about 10 jokes for my stand-up bit and another 20 for my "chat" with Jay afterward. It takes hours and hours of work at home and many nights onstage to get a *Tonight Show* routine the way I want it.

This girl was ugly. They used her in prisons to cure sex offenders.

Living as long as I have, you can't help but look back on life and wonder what does it all mean. Sometimes, I don't ever think I've made it. Even today, if I check into a hotel and a bellman picks up my suitcase, I feel awkward. *I feel like I should be taking the bags.* I guess I feel like I'm one of the masses. Maybe people know t<u>ha</u>t.

I've been broke most of my life. For years I was picking

RIGHT: Celebrating my eightieth birthday, on The Tonight Show. *Jay and Jim totally surprised me.*

up the phone and acting surprised. "The check came back? Oh!"

Every day now when I get up in the morning, I read the obituaries.

The obituaries have been very entertaining. Often you will read about the lives of some fascinating people.

I read about Suzanne Bloch, a musician and teacher. She was a class act, respected by all. Suzanne often played chamber music with noted scientists, including Albert Einstein.

Einstein turned out to be very difficult to work with. Suzanne would give the downbeat, but Einstein always came in late. Each time they had to go back to the beginning. Finally in exasperation, she turned on him and said, "Mr. Einstein, can't you count?"

But I can count and I know my days are numbered. I can picture my own funeral, the things that would be said:

> *We are here today to bid farewell to Rodney Dangerfield.*
> *A good husband, a good father, and a very good tipper.*
> *A man who cared about the homeless. He was always looking for a girl who needs a room.*

LEFT: *Here's me and Joan between operations.*

A man who always loved his neighbor—if she was easy.
Farewell, Rodney. We know you'll be in good hands—your own.

I tell ya one thing, though, I'm not about to die anytime soon. There are too many people out there who owe me money.

I can accept getting older. I can even accept getting old, but dying? Man, that's a tough one to accept. As my friend Joe Ancis used to say, "Who made this contract?"

Life's a short trip. You'll find out.

You were seventeen yesterday. You'll be fifty tomorrow. Life is tough, are you kiddin'? What do you think life is? Moonlight and canoes? That's not life. That's in the movies.

Life is fear and tension and worry and disappointments.

Life. I'll tell ya what life is. *Life is having a mother-in-law who sucks and a wife who don't.* That's what life is.

RIGHT: *My grandson, Joshua, has a great sense of humor. I just told him a joke. Look how he broke up.*

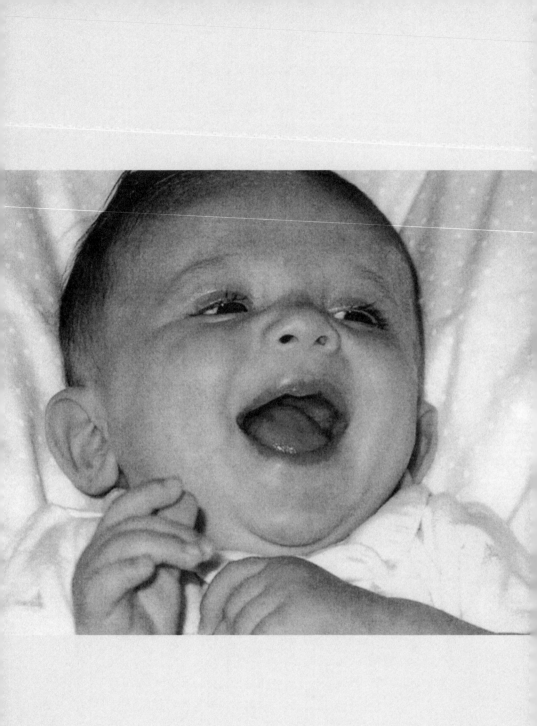

Acknowledgments

W ho should I thank first? I think it should be the guy who wrote the foreword to this book, Jim Carrey. Thank you, Jim, and thanks for being a real friend.

I also want to thank Chris Calhoun and the folks at Sterling Lord for their support and Bob Roe and Mic Kleber for their assistance.

This book also benefited from the talents and professionalism of the people at HarperCollins, especially my editor, David Hirshey.

Here's thanks to my bookie for being so patient.

I also want to take this opportunity to thank the following people for their friendship and kindnesses over the years:

David Permut, Harry Basil, Bob Saget, Billy Tragesser, Adam Sandler, Dr. Neal ElAttrache, Louie Anderson, Oliver Stone, Harland Williams, Dr. Neil Martin, Johnny Carson, Chris Albrecht, Paul Rodriguez, Jerry Stiller, Lenny Clarke, Anne Meara, Tim Allen, Anthony Bevacqua, Chris Albrecht, Joseph Merhi, Dr. Bruce Edwards, Marty Belafsky, Dom Irerra, Charlie Burke, Smokey Child, Tony Bennett, Merv Griffin, Warren Cowan, Richard Sturm, Dr. Grace Sun, Kevin Sasaki, Larry Shire, Dennis Arfa, Harold Ramis, Dr. Jamie Moriguchi, Robert Davi, Michael Bolton, and Jay Leno.

I want to thank the Chinese restaurant that delivers late.

And a big thanks to my wife, Joan; my son, Brian; my daughter, Melanie; her husband, David; and my grandson, Joshua.

Photograph Credits

Photograph Credits

Page 143: Courtesy of Dangerfield's, New York

Page 149: Courtesy of Dangerfield's, New York

Page 164: Courtesy of Dangerfield's, New York

Page 194: © Orion Pictures, Courtesy Everett Collection

Page 197: © Orion Pictures, Courtesy Everett Collection

Page 198: © Orion Pictures, Courtesy Everett Collection

Page 206: © Orion Pictures, Courtesy Everett Collection

Page 209: © Warner Bros., Courtesy Everett Collection

Page 250: © Orion Pictures, Courtesy Everett Collection

All other photographs courtesy of the collection of
Rodney Dangerfield.